Radu Precup
Ordinary Differential Equations

Also of Interest

Linear and Semilinear Partial Differential Equations. An Introduction
Radu Precup, 2012
ISBN 978-3-11-026904-8, e-ISBN (PDF) 978-3-11-026905-5

Elements of Partial Differential Equations
Pavel Drábek, Gabriela Holubová, 2014
ISBN 978-3-11-031665-0, e-ISBN (PDF) 978-3-11-031667-4,
e-ISBN (EPUB) 978-3-11-037404-9

Complex Analysis. A Functional Analytic Approach
Friedrich Haslinger, 2017
ISBN 978-3-11-041723-4, e-ISBN (PDF) 978-3-11-041724-1,
e-ISBN (EPUB) 978-3-11-042615-1

Numerical Analysis of Stochastic Processes
Wolf-Jürgen Beyn, Raphael Kruse, 2018
ISBN 978-3-11-044337-0, e-ISBN (PDF) 978-3-11-044338-7,
e-ISBN (EPUB) 978-3-11-043555-9

Nonlinear Equations with Small Parameter.
Volume I: Oscillations and Resonances
Sergey G. Glebov, Oleg M. Kiselev, Nikolai N. Tarkhanov, 2017
ISBN 978-3-11-033554-5, e-ISBN (PDF) 978-3-11-033568-2,
e-ISBN (EPUB) 978-3-11-038272-3

Nonlinear Equations with Small Parameter.
Volume II: Waves and Boundary Problems
Sergey G. Glebov, Oleg M. Kiselev, Nikolai N. Tarkhanov, 2018
ISBN 978-3-11-053383-5, e-ISBN (PDF) 978-3-11-053497-9,
e-ISBN (EPUB) 978-3-11-053390-3

Radu Precup

Ordinary Differential Equations

―

Example-driven, Including Maple Code

DE GRUYTER

Mathematics Subject Classification 2010
34-01

Author
Radu Precup
Babeş–Bolyai University
Department of Mathematics
Str. Mihail Kogălniceanu 1
400084 CLUJ-NAPOCA
Romania

ISBN 978-3-11-044742-2
e-ISBN (PDF) 978-3-11-044744-6
e-ISBN (EPUB) 978-3-11-044750-7

Library of Congress Cataloging-in-Publication Data
A CIP catalog record for this book has been applied for at the Library of Congress.

Bibliographic information published by the Deutsche Nationalbibliothek
The Deutsche Nationalbibliothek lists this publication in the Deutsche Nationalbibliografie; detailed bibliographic data are available on the Internet at http://dnb.dnb.de.

© 2018 Walter de Gruyter GmbH, Berlin/Boston
Cover image: Wavebreakmedia Ltd./Wavebreak Media/thinkstock
Typesetting: le-tex publishing services GmbH, Leipzig
Printing and binding: CPI books GmbH, Leck
♾ Printed on acid-free paper
Printed in Germany

www.degruyter.com

Non multa, sed multum

Preface

There are two main approaches to an introductory course in ordinary differential equations (differential equations, or ODEs for short). One is the classical treatment based on complete proofs and on the most general statements of the results. The other one, preferred by those readers interested in applications and mathematical modeling, covers difficult reasoning and focuses on the understanding of notions and methods by examples and exercises. Our approach is placed between them. On the one hand, by simple models arising from physics, biology and other areas, we try to provide sufficient motivation for the study of theory and for rigorous reasoning, and on the other hand, we constrain ourselves to a careful selection of material and a brief and accessible presentation.

Compared to other introductory books on ODEs, our option is to place the more delicate study of the Cauchy problem for first-order equations and systems at the end of the theoretical part, in such a way that the familiarity with differential equations gained until that stage makes the understanding of this theme easier. This option forces us to give a direct proof of the existence and uniqueness of the solution of the Cauchy problem for linear differential systems with constant coefficients. We think that the best way to do this is is by showing the parallelism between the first-order linear equation and the first-order linear systems, which formally have the same representation: scalar in the first case, and vectorial in the second case.

The book is structured in three parts: theory, applications, and computer simulation. Thus the text can be used as a first course in ODEs, as a collection of exercises and problems for tutorials, and as an introduction to scientific computation and simulation. A few projects are included in Part II as an invitation to those readers who wish, starting from an initial plan and bibliography, to test their ability to produce coherent material on a given theme based on free documentation and advanced study.

A number of special themes such as the Poincaré–Bendixson theorem, bifurcation problems, optimization of differential systems and control, stable manifolds, and vectorial methods for the treatment of systems, are briefly mentioned in the book in order to give the reader the option of further study in one of these modern directions.

The text is intended for students who desire an easy, quick and concise introduction to ODEs. It also addresses all those who are interested in mathematical modeling with ordinary differential equations, and in computer simulation. Additional theory, applications and exercises, as well as alternative presentations of the material, can be found by consulting other texts on ODEs, such as those in the Bibliography of this book.

The book is an improved version of a previous work published in Romanian and based on the lecture notes of a course on differential equations that I have been teaching for several years at the Faculty of Mathematics and Computer Science of Babeș–Bolyai University of Cluj-Napoca. I would like to thank Renata Bunoiu from the University of Metz, and my colleagues Adriana Buică and Ioan A. Rus, for careful reading of the manuscript and for their valuable remarks and suggestions. I also thank Marcel-Adrian Şerban for his valuable assistance in Part III.

June 2017											Radu Precup

Contents

Preface —— VII

Part I: Theory

Chapter 1 First-Order Differential Equations —— 3
1.1 Preliminaries —— 3
1.2 Classes of First-Order Differential Equations —— 6
1.2.1 Differential Equations with Separable Variables —— 6
1.2.2 Differential Equations of Homogeneous Type —— 7
1.2.3 First-Order Linear Differential Equations —— 8
1.2.4 Bernoulli Equations —— 13
1.2.5 Riccati Equations —— 14
1.3 Mathematical Modeling with First-Order Differential Equations —— 15
1.3.1 Radioactive Decay —— 15
1.3.2 Newton's Law of Heat Transfer —— 16
1.3.3 Chemical Reactions —— 17
1.3.4 Population Growth of a Single Species —— 18
1.3.5 The Gompertz Equation —— 20

Chapter 2 Linear Differential Systems —— 22
2.1 Preliminaries —— 22
2.2 Mathematical Modeling with Linear Differential Systems —— 23
2.3 Matrix Notation for Systems —— 26
2.4 Superposition Principle for Linear Systems —— 27
2.5 Linear Differential Systems with Constant Coefficients —— 28
2.5.1 The General Solution —— 28
2.5.2 Structure of Solution Set for Homogeneous Linear Systems —— 31
2.5.3 The Concept of Fundamental Matrix —— 32
2.5.4 Method of Eigenvalues and Eigenvectors —— 35
2.6 Method of Variation of Parameters —— 40
2.7 Higher-Dimensional Linear Systems —— 42
2.8 Use of the Jordan Canonical Form of a Matrix —— 44
2.9 Dynamic Aspects of Differential Systems —— 46
2.10 Preliminaries of Stability —— 52

Chapter 3 Second-Order Differential Equations —— 53
3.1 Newton's Second Law of Motion —— 53
3.2 Reduction of Order —— 54

3.3	Equivalence to a First-Order System —— 57	
3.4	The Method of Elimination —— 58	
3.5	Linear Second-Order Differential Equations —— 59	
3.5.1	The Solution Set —— 59	
3.5.2	Homogeneous Linear Equations with Constant Coefficients —— 59	
3.5.3	Variation of Parameters Method —— 61	
3.5.4	The Method of Undetermined Coefficients —— 62	
3.5.5	Euler Equations —— 64	
3.6	Boundary Value Problems —— 66	
3.7	Higher-Order Linear Differential Equations —— 69	

Chapter 4 Nonlinear Differential Equations —— 72
- 4.1 Mathematical Models Expressed by Nonlinear Systems —— 72
- 4.1.1 The Lotka–Volterra Model —— 72
- 4.1.2 The SIR Epidemic Model —— 73
- 4.1.3 An Immunological Model —— 74
- 4.1.4 A Model in Hematology —— 75
- 4.2 Gronwall's Inequality —— 75
- 4.3 Uniqueness of Solutions for the Cauchy Problem —— 77
- 4.4 Continuous Dependence of Solutions on the Initial Values —— 81
- 4.5 The Cauchy Problem for Systems —— 83
- 4.6 The Cauchy Problem for Higher-Order Equations —— 84
- 4.7 Periodic Solutions —— 85
- 4.8 Picard's Method of Successive Approximations —— 88
- 4.8.1 Picard's Iteration —— 88
- 4.8.2 The Interval of Picard's Iteration —— 91
- 4.8.3 Convergence of Picard's Iteration —— 92
- 4.9 Existence of Solutions for the Cauchy Problem —— 93

Chapter 5 Stability of Solutions —— 97
- 5.1 The Notion of a Stable Solution —— 97
- 5.2 Stability of Linear Systems —— 99
- 5.3 Stability of Linear Systems with Constant Coefficients —— 100
- 5.4 Stability of Solutions of Nonlinear Systems —— 101
- 5.5 Method of Lyapunov Functions —— 106
- 5.6 Globally Asymptotically Stable Systems —— 111

Chapter 6 Differential Systems with Control Parameters —— 113
- 6.1 Bifurcations —— 113
- 6.2 Hopf Bifurcations —— 115
- 6.3 Optimization of Differential Systems —— 119
- 6.4 Dynamic Optimization of Differential Systems —— 122

Part II: Exercises

Seminar 1	**Classes of First-Order Differential Equations** —— 127	
1.1	Solved Exercises —— 127	
1.2	Proposed Exercises —— 129	
1.3	Solutions —— 130	
1.4	Project: Problems of Geometry that Lead to Differential Equations —— 131	

Seminar 2	**Mathematical Modeling with Differential Equations** —— 133	
2.1	Solved Exercises —— 133	
2.2	Proposed Exercises —— 135	
2.3	Hints and Answers —— 136	
2.4	Project: Influence of External Actions over the Evolution of Some Processes —— 136	

Seminar 3	**Linear Differential Systems** —— 139	
3.1	Solved Exercises —— 139	
3.2	Proposed Exercises —— 143	
3.3	Hints and Solutions —— 145	
3.4	Project: Mathematical Models Represented by Linear Differential Systems —— 146	

Seminar 4	**Second-Order Differential Equations** —— 148	
4.1	Solved Exercises —— 148	
4.2	Proposed Exercises —— 150	
4.3	Solutions —— 151	
4.4	Project: Boundary Value Problems for Second-Order Differential Equations —— 152	

Seminar 5	**Gronwall's Inequality** —— 154	
5.1	Solved Exercises —— 154	
5.2	Proposed Exercises —— 157	
5.3	Hints and Solutions —— 157	
5.4	Project: Integral and Differential Inequalities —— 158	

Seminar 6	**Method of Successive Approximations** —— 163	
6.1	Solved Exercises —— 163	
6.2	Proposed Exercises —— 164	

6.3	Hints and Solutions —— 165
6.4	Project: The Vectorial Method for the Treatment of Nonlinear Differential Systems —— 165

Seminar 7 Stability of Solutions —— 170
7.1	Solved Exercises —— 170
7.2	Proposed Exercises —— 172
7.3	Hints and Solutions —— 173
7.4	Project: Stable and Unstable Invariant Manifolds —— 173

Part III: Maple Code

Lab 1 Introduction to Maple —— 179
1.1	Numerical Calculus —— 179
1.2	Symbolic Calculus —— 181

Lab 2 Differential Equations with Maple —— 185
2.1	The DEtools Package —— 185
2.2	Working Themes —— 186

Lab 3 Linear Differential Systems —— 188
3.1	The linalg Package —— 188
3.2	Linear Differential Systems —— 188
3.3	Working Themes —— 190

Lab 4 Second-Order Differential Equations —— 191
4.1	Spring-Mass Oscillator Equation with Maple —— 191
4.2	Boundary Value Problems with Maple —— 193
4.3	Working Themes —— 193

Lab 5 Nonlinear Differential Systems —— 195
5.1	The Lotka–Volterra System —— 195
5.2	A Model from Hematology —— 196
5.3	Working Themes —— 197

Lab 6 Numerical Computation of Solutions —— 198
6.1	Initial Value Problems —— 198
6.2	Boundary Value Problems —— 199
6.3	Working Themes —— 202

Lab 7	**Writing Custom Maple Programs** —— 204
7.1	Method of Successive Approximations —— 204
7.2	Euler's Method —— 206
7.3	The Shooting Method —— 208
7.4	Working Themes —— 211
Lab 8	**Differential Systems with Control Parameters** —— 212
8.1	Bifurcations —— 212
8.2	Optimization with Maple —— 213
8.3	Working Themes —— 215

Bibliography —— 217

Index —— 219

Part I: **Theory**

Chapter 1 First-Order Differential Equations

1.1 Preliminaries

Roughly speaking, an *ordinary differential equation* (frequently called 'differential equation', or 'ODE') is an equation that connects an unknown function of one variable to some of its derivatives. The order of the highest derivative that occurs is called the *order* of the equation. Thus, if we denote the independent variable by t and the function of variable t, by x then the *general form* of an n-th order differential equation is

$$F\left(t, x, x', \ldots, x^{(n)}\right) = 0. \tag{1.1}$$

By a *solution* of the equation (1.1) we mean a function $x \in C^n(J)$ defined on some interval J of real numbers (the domain of the solution), which satisfies the differential equation for all values of t from J, i.e.

$$F\left(t, x(t), x'(t), \ldots, x^{(n)}(t)\right) = 0 \quad \text{for every } t \in J.$$

Different solutions of the same equation can have different domains. According to our definition, if $x \in C^n(J)$ is a solution, then its restriction to any subinterval of J is also a solution. We say that the domain of a solution is *maximal* if that solution is not the restriction of any solution of a larger domain, or, equivalently, if it cannot be continuable. The solutions whose domains are maximal are called *saturated solutions*.

For example, the functions $x_1 = \ln t$ and $x_2 = \ln(t - 1)$ are both solutions of the first-order differential equation $x' - e^{-x} = 0$, and their maximal domains are $(0, +\infty)$ and $(1, +\infty)$, respectively. More generally, any function of the form $x = \ln(t - c)$, where c is a real constant, is a solution of this equation and its maximal domain is $(c, +\infty)$.

If from relation (1.1), the highest derivative $x^{(n)}$ can be expressed in terms of t and of the lower derivatives $x, x', \ldots, x^{(n-1)}$, then one obtains an equation of the form

$$x^{(n)} = f\left(t, x, x', \ldots, x^{(n-1)}\right). \tag{1.2}$$

This is the *normal form* of n-th order differential equations.

For example, the equation $x - t + \ln x' = 0$ can be put under the normal form $x' = e^{t-x}$.

The simplest differential equation of order n arises if the function f does not depend on the unknown function, so that the equation is

$$x^{(n)} = h(t),$$

where $h \in C(J)$. Its solutions are obtained by successive n times integrations of h. After each integration an arbitrary constant will occur. For example, for $n = 2$, if we fix a point $t_0 \in J$, from $x'' = h(t)$ we first obtain

$$x'(t) = \int_{t_0}^{t} h(s)\, ds + C_1$$

and then all the solutions of the equation, namely

$$x(t) = \int_{t_0}^{t}\int_{t_0}^{\tau} h(s)\, ds\, d\tau + C_1 t + C_2, \tag{1.3}$$

where C_1, C_2 are arbitrary constants. Thus, the set of solutions depends on a set of arbitrary constants whose number is equal to the order of the equation. We say that (1.3) is the general solution of the equation $x'' = h(t)$. More generally, the solution of an ordinary differential equation of order n that contains n arbitrary constants is called the *general solution* of that equation. Now if we wish to select out from (1.3) a particular solution, then we need to fix the a priori values of x and x' at t_0. Let them be α_0 and α_1, then we obtain $C_1 = \alpha_1$ and $C_2 = \alpha_0 - \alpha_1 t_0$. We say that we solved the initial value problem, or the Cauchy[1] problem

$$\begin{cases} x'' = h(t) \\ x(t_0) = \alpha_0, \quad x'(t_0) = \alpha_1. \end{cases}$$

Extrapolating and supposing that something similar happens in the general case, we can formulate the *initial value problem*, or the *Cauchy problem* for the equation (1.2), as follows:

$$\begin{cases} x^{(n)} = f\left(t, x, x', \ldots, x^{(n-1)}\right) \\ x^{(k)}(t_0) = \alpha_k, \quad k = 0, 1, \ldots, n-1. \end{cases}$$

In this problem, t_0 is the *initial point*, while α_k, $k = 0, 1, \ldots, n-1$ are the *initial values* of the functions $x, x', \ldots, x^{(n-1)}$.

If in (1.1) and (1.2), the expressions of F and f do not explicitly depend on t (but only implicitly by means of the function x and of its derivatives), we say that the equation is *autonomous*. Thus, the general form of autonomous n-th order differential equations is

$$F\left(x, x', x'', \ldots, x^{(n)}\right) = 0,$$

and their normal form is

$$x^{(n)} = f\left(x, x', \ldots, x^{(n-1)}\right).$$

Otherwise, the equation is said to be *nonautonomous*.

---

[1] Augustin Louis Cauchy (1789–1857)

For example, the equation $x' - e^{-x} = 0$ is autonomous, but the equation $x - t + \ln x' = 0$ is nonautonomous.

In another classification of the differential equations, the distinction is made between linear and nonlinear equations. Thus, the equation (1.1) is *linear* if F depends linearly on $x, x', \ldots, x^{(n)}$, that is, it has the form

$$a_0(t)x^{(n)} + a_1(t)x^{(n-1)} + \ldots + a_n(t)x = h(t). \tag{1.4}$$

The *coefficients* $a_0, a_1, \ldots, a_n$ and the *right-hand side* h, also called the *nonhomogeneous part*, are functions of variable t on a given interval J, and a_0 is not identically zero on J. If h is identically zero, then the equation is called *homogeneous*. Otherwise, the equation is called *nonhomogeneous*. If the coefficients $a_0, a_1, \ldots, a_n$ are constant, then (1.4) is called a *constant coefficient differential equation*. Thus, the general form of a constant coefficient equation of order n is

$$x^{(n)} + a_1 x^{(n-1)} + \ldots + a_n x = h(t),$$

where $a_1, a_2, \ldots, a_n$ are constants and h is a given function.

For example, the equation $x'' + tx' + x = 1$ is linear but nonhomogeneous, and is not a constant coefficient equation; the equation $x'' - 3x' + x = t + 1$ is a nonhomogeneous linear constant coefficient equation; while the equation $x'' - 3x' + x = 0$ is a homogeneous linear constant coefficient equation.

Also, it is clear that a linear equation is autonomous if and only if it is a constant coefficient equation with a constant right-hand side.

In what follows we shall mainly consider first- and second-order differential equations. Thus, for $n = 1$, the first-order differential equations have the general form

$$F(t, x, x') = 0,$$

and the normal form

$$x' = f(t, x).$$

Notice that the notations t and x used to denote the independent variable and the unknown function are not standard. Instead, one can use some other characters, such as t and u, or x and y, or x and u, etc. The choice of notations is a matter of taste, or, as we shall see, it can be suggested by the nature (geometrical, physical, chemical, biological, etc.) of the processes or problems mathematically modeled/expressed by differential equations. In this book, the notations t, x will be preferentially but not exclusively used.

In the next section, we will discuss some classes of first-order differential equations that can be solved by direct integration.

1.2 Classes of First-Order Differential Equations

1.2.1 Differential Equations with Separable Variables

A first-order differential equation is said to have *separable variables,* if the function $f(t, x)$ is a product of a continuous function of t and a continuous function of x, that is, if it has the form
$$x' = g(t)h(x) . \qquad (1.5)$$
To solve it, we separate the variables t and x by putting the equation into the form
$$\frac{x'}{h(x)} = g(t) .$$
Now observe that the left-hand side is the derivative with respect to t of the function $H \circ x$, where H is a primitive of the function $1/h$. Then, integrating both sides with respect to t gives
$$H(x) = \int g(t) \, dt .$$
Solving for x, assuming that H is invertible, yields the solution
$$x = H^{-1}\left(\int g(t) \, dt\right) . \qquad (1.6)$$
Since $\int g(t) \, dt$ is a family of functions involving an arbitrary integration constant, we may conclude that the solution set depends on an arbitrary constant. Thus (1.6) is the general solution of (1.5). Since we divided by $h(x)$, the calculation assumed that $h(x) \neq 0$. Note that any zero of h as a constant function satisfies the differential equation (1.5) but is not included in the general solution (1.6). Therefore, the solutions of (1.5) consist of (1.6) and all constants k with $h(k) = 0$.

Example 1.1. Solve the differential equation
$$x' = \frac{1}{t}x \quad \text{for } t > 0 .$$

Solution. The equation has separable variables; here $g(t) = 1/t$ and $h(x) = x$. Separating variables gives
$$\frac{x'}{x} = \frac{1}{t} .$$
Integration with respect to t yields $\ln|x| = \ln t + c$, where c is an arbitrary constant. It is useful to represent this constant in the form $c = \ln C$, where $C > 0$. Then $\ln|x| = \ln t + \ln C$, and $|x| = Ct$. If we eliminate the absolute value sign, then we obtain $x = Ct$, with $C \in \mathbb{R} \setminus \{0\}$. Notice that the value $C = 0$ is also acceptable since $h(0) = 0$. Thus, the general solution of the equation is $x = Ct$, where C is an arbitrary constant. If we now consider the Cauchy problem with the initial condition $x(1) = 2$, then the constant C can be determined, namely $C = 2$, and the unique solution of the equation that satisfies the initial condition $x(1) = 2$ is $x = 2t$.

Example 1.2. Solve the Cauchy problem

$$x' = 2t(x^2 + 1), \quad x(0) = 1.$$

Solution. Separating variables gives

$$\frac{x'}{x^2 + 1} = 2t,$$

and by integration with respect to t, $\arctan x = t^2 + C$. Solving for x yields $x = \tan(t^2 + C)$ depending on an arbitrary constant C, which is the general solution of the equation. Now, making use of the initial condition, we deduce that $\tan C = 1$, so $C = \pi/4 + k\pi$ ($k \in \mathbb{Z}$). Substituting C into the solution formula, and taking into account that $k\pi$ is a period of the tangent function, yields the unique solution of the Cauchy problem: $x = \tan(t^2 + \pi/4)$.

1.2.2 Differential Equations of Homogeneous Type

A differential equation that has the form

$$x' = g\left(\frac{x}{t}\right), \tag{1.7}$$

where g is a continuous function of a single variable, is said to be of homogeneous type. To solve (1.7), we make the substitution $x = ty$. Then by the product rule of the derivative, $y + ty' = g(y)$, or

$$y' = \frac{1}{t}(g(y) - y). \tag{1.8}$$

Equation (1.8) has separable variables and so it can be solved as above. Let $y = \phi(t, C)$ be its general solution depending on an arbitrary constant C. Then the general solution of (1.7) is $x = t\phi(t, C)$. Therefore, any equation of homogeneous type can be reduced to a differential equation with separable variables by the substitution $x = ty$.

Example 1.3. Find the solutions of the equation

$$x' = -\left(\frac{x}{t}\right)^2 + \frac{x}{t} \quad \text{for } t > 0.$$

Solution. The equation is of homogeneous type with the form (1.7) with $g(\tau) = -\tau^2 + \tau$. Making the substitution $x = ty$ gives $y + ty' = -y^2 + y$, or

$$ty' = -y^2. \tag{1.9}$$

We separate the variables t and y to obtain

$$-\frac{y'}{y^2} = \frac{1}{t}.$$

Integrating with respect to t yields $1/y = \ln t + C$, which leads to $y = 1/(\ln t + C)$. Finally we obtain the general solution

$$x = \frac{t}{\ln t + C}. \tag{1.10}$$

Note that $x = 0$ is also a solution that is not included in the general solution (1.10). It corresponds to the solution $y = 0$ of (1.9), lost in the process of separation of variables.

Example 1.4. Solve the Cauchy problem

$$y' = \frac{y}{x+y} \quad (x, y > 0), \quad y(1) = 1.$$

Solution. Here the variables are x and y instead of the previous ones, t and x. Since the differential equation can be written as

$$y' = \frac{\frac{y}{x}}{1 + \frac{y}{x}},$$

it is of homogeneous type. With the change of variable $y = xz$, we have $z + xz' = z/(1+z)$, which is the separable equation

$$xz' = \frac{-z^2}{1+z}. \tag{1.11}$$

Furthermore, $z'(1+z)z^{-2} = -1/x$, or

$$z'\left(\frac{1}{z^2} + \frac{1}{z}\right) = -\frac{1}{x}.$$

Integrating with respect to x gives

$$-\frac{1}{z} + \ln z = -\ln x + C. \tag{1.12}$$

Since from equation (1.12) we cannot obtain z explicitly in terms of x, we are content with saying that (1.12) gives the general solution of (1.11) in the implicit form, or that (1.12) is the *implicit general solution* of (1.11). Next, the initial condition requires $x = 1$ to have $y = 1$ and so $z = 1$. Letting $x = 1$ and $z = 1$ in (1.12), we find that $C = -1$. Thus, the solution of the Cauchy problem is the function $y = xz$, where z is the function of variable x given implicitly by the relation

$$-\frac{1}{z} + \ln z = -\ln x - 1.$$

1.2.3 First-Order Linear Differential Equations

A differential equation that can be expressed as

$$x' = a(t)x + b(t), \tag{1.13}$$

where a and b are given continuous functions defined on some interval J, is called a *linear differential equation of the first order*. If $b(t) = 0$ for every $t \in J$, then the equation is called *homogeneous*.

Integrating Factor

For simplicity we shall present the method of integrating factors for the linear differential equation with a constant coefficient $a(t) = a$,

$$x' = ax + b(t). \tag{1.14}$$

The technique for solving (1.14) is to look for a function $h(t)$ so that after multiplication by $h(t)$, the expression containing the unknown function x, namely $h(t)x'(t) - ah(t)x(t)$, represents the derivative of some product function. The reader can see that such a function $h(t)$, called an *integrating factor*, is

$$h(t) = e^{-ta}.$$

With this choice, since $h'(t) = -ah(t)$, one has

$$h(t)x'(t) - ah(t)x(t) = \frac{d}{dt}(h(t)x(t)),$$

and so, after multiplication by $h(t)$, (1.14) becomes

$$\frac{d}{dt}(h(t)x(t)) = h(t)b(t).$$

Integration from a given point $t_0 \in J$ to an arbitrary t gives

$$h(t)x(t) = C + \int_{t_0}^{t} h(s)b(s)\,ds.$$

Now dividing by $h(t) = e^{-ta}$ gives the general solution

$$x = e^{ta}C + \int_{t_0}^{t} e^{(t-s)a}b(s)\,ds, \tag{1.15}$$

where C is an arbitrary constant. Moreover, if we have the initial condition $x(t_0) = x_0$, then taking $t = t_0$ and $x = x_0$ in (1.15), we obtain $C = e^{-t_0 a}x_0$, so the unique solution of the Cauchy problem is

$$x = e^{(t-t_0)a}x_0 + \int_{t_0}^{t} e^{(t-s)a}b(s)\,ds.$$

As we shall see later in Chapter 2, the method that we have just presented can be extended for solving systems of first-order linear differential equations with constant coefficients.

The technique using an integrating factor also applies to equation (1.13). In this case an integrating factor is

$$h(t) = e^{-\int_{t_0}^{t} a(\tau)\,d\tau}, \tag{1.16}$$

for some $t_0 \in J$. Following the same arguments as above, we obtain the general solution of the equation (1.13),

$$x = e^{\int_{t_0}^t a(\tau)\, d\tau} C + \int_{t_0}^t e^{\int_s^t a(\tau)\, d\tau} b(s)\, ds, \qquad (1.17)$$

depending on an arbitrary constant C, and the unique solution of the Cauchy problem with the initial condition $x(t_0) = x_0$,

$$x = e^{\int_{t_0}^t a(\tau)\, d\tau} x_0 + \int_{t_0}^t e^{\int_s^t a(\tau)\, d\tau} b(s)\, ds.$$

Notice that all saturated solutions of a linear differential equation have the same domain, namely the whole interval J on which the functions a and b are defined.

Example 1.5. Using the technique of an integrating factor, solve the differential equation

$$x' + \frac{1}{t} x = 3t \quad \text{for } t > 0. \qquad (1.18)$$

Solution. The equation is linear with $a(t) = -1/t$ and $b(t) = 3t$. If we choose $t_0 = 1$, then according to (1.16), the integrating factor is $h(t) = \exp(\ln t) = t$. Hence, multiply by t and find $tx' + x = 3t^2$, or $(tx)' = 3t^2$. Consequently, $tx = C + t^3$. Finally we obtain the general solution

$$x = \frac{C}{t} + t^2,$$

where C is an arbitrary constant.

The Superposition Principle

An alternative method of solving linear differential equations, which does requires neither an integrating factor nor the memorization of the representation formula (1.17) and, as we shall see, is applicable to all linear differential equations and systems, is based on the algebraic connection between the general solution of the nonhomogeneous equation (1.13) and the general solution of the associated homogeneous equation

$$x' = a(t)x. \qquad (1.19)$$

This connection is given by the following theorem expressing the so-called *superposition principle* for linear differential equations.

Theorem 1.6. *The general solution x of any linear differential equation (1.13) can be represented in the form*

$$x = x_h + x_p,$$

where x_h is the general solution of the associated homogeneous equation (1.19) and x_p is a particular solution of (1.13).

Proof. Let x_p be a particular solution of (1.13). If x is any solution of the equation (1.13), then from

$$x' = a(t)x + b(t)$$
$$x'_p = a(t)x_p + b(t)$$

by subtraction, we obtain

$$(x - x_p)' = a(t)(x - x_p) \,.$$

Thus the function $x_h := x - x_p$ is a solution of the homogeneous equation (1.19). Clearly, $x = x_h + x_p$. Conversely, if x_h is any solution of the homogeneous equation (1.19), then from

$$x'_h = a(t)x_h$$
$$x'_p = a(t)x_p + b(t)$$

by addition, we deduce that

$$(x_h + x_p)' = a(t)(x_h + x_p) + b(t) \,.$$

Hence the function $x := x_h + x_p$ is a solution of the equation (1.13). □

According to Theorem 1.6, to solve linear differential equations, we should follow three steps:
- Solve the associated homogeneous equation;
- Find a particular solution of the nonhomogeneous equation;
- Write down the general solution as $x = x_h + x_p$.

Step 1. The associated homogeneous equation (1.19) is separable. Separating variables gives $x'/x = a(t)$, and integrating yields

$$\ln|x| = \int_{t_0}^{t} a(\tau)\, d\tau + \ln C,$$

where C is an arbitrary positive constant. Exponentiation of both sides leads to

$$|x| = e^{\int_{t_0}^{t} a(\tau)\, d\tau} C,$$

and elimination of the absolute value sign gives the general solution of the homogeneous equation (1.19),

$$x_h = e^{\int_{t_0}^{t} a(\tau)\, d\tau} C, \tag{1.20}$$

where C is now any (positive, negative or zero) constant.

Step 2. A particular solution of the nonhomogeneous equation (1.13) can be obtained by a standard method called *variation of parameters*. The idea is to try a solution inspired by (1.20) where we let C vary with respect to t. Hence we look for a solution to (1.13) of the form

$$x_p = e^{\int_{t_0}^{t} a(\tau)\,d\tau} C(t) . \tag{1.21}$$

To find $C(t)$ it is enough to substitute the expression of x_p into the nonhomogeneous equation. This yields

$$a(t)e^{\int_{t_0}^{t} a(\tau)\,d\tau} C(t) + e^{\int_{t_0}^{t} a(\tau)\,d\tau} C'(t)$$
$$= a(t)e^{\int_{t_0}^{t} a(\tau)\,d\tau} C(t) + b(t) .$$

Here the terms containing $C(t)$ cancel giving

$$e^{\int_{t_0}^{t} a(\tau)\,d\tau} C'(t) = b(t) ,$$

or

$$C'(t) = e^{-\int_{t_0}^{t} a(\tau)\,d\tau} b(t) .$$

Thus, we may choose

$$C(t) = \int_{t_0}^{t} e^{-\int_{t_0}^{s} a(\tau)\,d\tau} b(s)\,ds .$$

With this choice, (1.21) becomes

$$x_p = \int_{t_0}^{t} e^{\int_{s}^{t} a(\tau)\,d\tau} b(s)\,ds . \tag{1.22}$$

Step 3. The general solution of the original equation is $x = x_h + x_p$, which according to (1.20) and (1.22) admits the representation (1.17).

Example 1.7. Solve the equation (1.18) using Theorem 1.6.

Solution. First consider the associated homogeneous equation $x' + x/t = 0$. Separating the variables and integrating with respect to t gives $\ln|x| = -\ln t + \ln C$, with $C > 0$. We deduce that $x_h = C/t$, where C is an arbitrary real constant. Next, look for a particular solution of the original equation, in the form $x_p = C(t)/t$. One has $x_p' = C'(t)/t - C(t)/t^2$. If we substitute x_p into the equation, then we find

$$C'(t)\frac{1}{t} - C(t)\frac{1}{t^2} + C(t)\frac{1}{t^2} = 3t .$$

Hence $C'(t) = 3t^2$, and we may choose $C(t) = t^3$. This gives $x_p = t^2$ and, finally, the general solution

$$x = \frac{C}{t} + t^2 ,$$

with $C \in \mathbb{R}$.

1.2.4 Bernoulli Equations

A differential equation of the form

$$x' = a(t)x + b(t)x^\alpha,$$

where α is any real number different from 0 and 1, is called a *Bernoulli*[2] *equation*. Note that for $\alpha = 0$ or $\alpha = 1$, the equation is linear. Let us name a famous special case of the Bernoulli differential equation, the logistic equation

$$x' = ax(1 - bx),$$

which will appear in Section 1.3.

The Bernoulli equation can always be reduced to a linear equation by changing the dependent variable from x to y via $y = x^{1-\alpha}$. We are led to this change of variable if we divide both sides of the equations by x^α to get

$$x^{-\alpha} x' = a(t) x^{1-\alpha} + b(t). \tag{1.23}$$

We then observe that the term $x^{-\alpha} x'$ comes from the differentiation of the function $x^{1-\alpha}$. Thus, (1.23) becomes

$$\frac{1}{1-\alpha} y' = a(t) y + b(t),$$

or

$$y' = (1 - \alpha) a(t) y + (1 - \alpha) b(t),$$

which is a linear equation.

Example 1.8. Making use of a suitable change of variable, show that the equation

$$x' + tx = e^{t^2} x^3$$

can be reduced to a linear equation and then find its solutions.

Solution. The equation is of Bernoulli type. Division by x^3 yields $x' x^{-3} + t x^{-2} = e^{t^2}$. Next we change the variable via $y := x^{-2}$. Since $y' = -2x^{-3} x'$, the equation becomes a linear equation,

$$y' - 2ty = -2e^{t^2}. \tag{1.24}$$

First we solve the associated homogeneous equation $y' - 2ty = 0$ and get $y_h = Ce^{t^2}$. Next, with the method of the variation of parameters, we find a particular solution of the nonhomogeneous equation, namely $y_p = -2te^{t^2}$. Thus, the general solution of (1.24) is

$$y = y_h + y_p = (C - 2t) e^{t^2}.$$

---

[2] Jacob Bernoulli (1654–1705)

Finally, coming back to x we obtain the solutions of the Bernoulli equation,

$$x = \pm\sqrt{\frac{1}{y}} = \pm\frac{e^{-\frac{t^2}{2}}}{\sqrt{C - 2t}},$$

where C is an arbitrary constant.

1.2.5 Riccati Equations

A differential equation of the form

$$x' = a(t)x + b(t)x^2 + c(t), \tag{1.25}$$

where a, b, c are continuous functions on an interval J, and c is not identically zero in J, is called a *Riccati*[3] *equation*. Notice that if $c \equiv 0$, then (1.25) is a Bernoulli equation. In general, the Riccati equations cannot be solved by elementary functions. However, if one solution is known, then all the other solutions can be found. Indeed, if $x = \varphi(t)$ is a solution of (1.25), then via the change of variable

$$y = x - \varphi,$$

we obtain

$$y' + \varphi' = a(t)y + a(t)\varphi + b(t)\left(y^2 + 2\varphi y + \varphi^2\right) + c(t). \tag{1.26}$$

Since φ is a solution of (1.25), i.e.

$$\varphi' = a(t)\varphi + b(t)\varphi^2 + c(t),$$

from (1.26) we derive the Bernoulli equation

$$y' = [a(t) + 2b(t)\varphi(t)]\, y + b(t)y^2,$$

which can be solved as shown before.

Thus, if by some method a solution φ of the Riccati equation is known, then by the change of variable $y = x - \varphi$, the Riccati equation is reduced to a Bernoulli equation.

Example 1.9. Solve the equation

$$x' = -2xe^t + x^2 + e^t + e^{2t}$$

if it is known that e^t is one of its solutions.

Solution. The equation is of Riccati type. By the change of variable $y = x - e^t$ the equation is reduced to the Bernoulli equation $y' = y^2$, which in fact is a separable

---

[3] Jacopo Riccati (1676–1754)

equation. Its general solution is $y = 1/(C-t)$. Thus the general solution of the original equation is

$$x = e^t + \frac{1}{C-t},$$

where C is an arbitrary constant.

1.3 Mathematical Modeling with First-Order Differential Equations

This section is intended to give the motivation for the study of differential equations as an integral part of applied mathematics, that is, of those mathematics directed towards real world applications. We shall present some simple examples of first-order differential equations arising from mathematical modeling of real processes from physics, chemistry and biology. This form of presentation will be carried out throughout this book by other models given by second-order differential equations and by systems of linear or nonlinear differential equations.

The fact that many laws of nature can be stated as differential equations is not surprising. Indeed, the state of a phenomenon is often described by some 'quantity' $x(t)$ that changes in time and whose rates of changing are given by its derivatives $x'(t)$, $x''(t)$, etc. For example, in mechanics, in many cases the meaning of $x(t)$, $x'(t)$ and $x''(t)$ is position, velocity and acceleration at time t; in biology, $x(t)$ may have the meaning of population as number of individuals or as a density, where $x'(t)$ gives the rate of population growth; and in chemistry $x(t)$ could represent the quantity or concentration of a chemical substance. By connecting a function $x(t)$ with some of its derivatives, an ordinary differential equation can show how the 'quantity' or the state of a process changes in time. The solutions of the differential equation as a model (the general solution) can be physically understood as set of all possible evolutions of the process; the initial condition can be seen as the initial state of the process, while the unique solution of the Cauchy problem should describe the real evolution of the process, completely determined by the initial state, in accordance with the physical causal determinism.

1.3.1 Radioactive Decay

For a given radioactive material, let $x(t)$ be the number of atoms at time t. According to the experimental physical law, the rate of decay for the radioactive material is proportional to the amount of the radioactive substance. Mathematically, this is expressed by $x'(t) = -kx(t)$, which corresponds to the differential equation

$$x' = -kx, \qquad (1.27)$$

where k is a positive proportionality constant characterizing each radioactive material, called the *decay constant*. The minus sign is introduced since radioactivity diminishes the number of active atoms, so that $x(t)$ is a decreasing function and its derivative $x'(t)$ has to be negative. As a first-order homogeneous equation, (1.27) can be solved by separation of variables. Its general solution depending on an arbitrary constant is

$$x(t) = e^{-kt}C$$

and shows that the amount of radioactive material decreases exponentially to zero as $t \to +\infty$. If at initial time t_0 the amount x_0 of radioactive material is known, then $x_0 = e^{-kt_0}C$, and so the constant C is uniquely determined to be $C = e^{kt_0}x_0$. Consequently, the amount of radioactive material at any time t is given by

$$x(t) = e^{-k(t-t_0)}x_0. \tag{1.28}$$

Mathematically, this is the solution of the Cauchy problem with the initial condition $x(t_0) = x_0$, and physically, it is in accordance with the principle of scientific determinism requiring the evolution of a process to be uniquely determined by its initial state.

A natural question is how the decay constant k of a given material can be determined. To give an answer, suppose that the amount of radioactive substance is estimated at two moments of time t_0 and t_1 to be x_0 and x_1, respectively. Then, from

$$x_1 = e^{-k(t_1-t_0)}x_0,$$

by taking the logarithm, we obtain the constant k as

$$k = -\frac{1}{t_1 - t_0} \ln \frac{x_1}{x_0}.$$

A more commonly used parameter is the *half-life* denoted T and representing the length of time it takes the material to decay to half its initial amount. If in (1.28) we put $t = t_0 + T$ and $x(t_0 + T) = x_0/2$ (half the initial amount), then we find $x_0/2 = e^{-kT}x_0$, and after taking the logarithm we get the formula

$$T = \frac{\ln 2}{k},$$

connecting the two constants k and T.

1.3.2 Newton's Law of Heat Transfer

Suppose that an object of uniform temperature x_0 is placed at time t_0 in an environment of constant temperature x_e. If $x_e < x_0$, then the temperature of the object decreases (the object cools down), while if $x_e > x_0$, then it increases (the object heats up). Newton's[4] law of heat transfer states that the rate of change of the temperature of

---

[4] Isaac Newton (1643–1727)

the object is proportional to the difference between the temperature of the object and the environmental temperature. Obviously, in this model, it is assumed that at any time t, the temperature is the same throughout the inside of the object, and its change is instantaneous. If $x(t)$ is the temperature of the object at time t, then *Newton's law of heat transfer* becomes

$$x' = -k(x - x_e), \tag{1.29}$$

where k is a positive proportionality constant called the *heat transfer coefficient*, which depends on the thermal properties of the object and environment. Equation (1.29) is a first-order linear differential equation, which can be solved by the integrating factor technique, or using the superposition principle, Theorem 1.6. The details are left as an exercise to the reader. Under the initial condition $x(t_0) = x_0$, the unique solution of the equation is

$$x(t) = x_e + (x_0 - x_e)e^{-k(t-t_0)}.$$

Therefore, the temperature of the object follows an exponential law. In addition, since $x(t) \to x_e$ as $t \to +\infty$, the temperature of the object approaches the environmental temperature as time increases. Thus the mathematical result confirms the physical expectation.

The reader is advised that to find a method for the determination of the heat transfer coefficient k, they should be inspired by the previous example.

1.3.3 Chemical Reactions

Consider a reversible chemical reaction $X \rightleftarrows Y$ consisting of the transformation of compound X into compound Y and vice versa, and assume the conservation of the total concentration of X and Y. Denote by x_0, y_0 the initial concentrations of the two compounds at time $t_0 = 0$ and by $x(t), y(t)$ their concentrations at any time t. By our assumption,

$$x(t) + y(t) = x_0 + y_0 =: m \quad \text{(constant)}.$$

If a and b are the transformation rates for the reactions $X \to Y$ and $Y \to X$ respectively, then we may write the balance equation for X,

$$x' = -ax + by.$$

Since $y = m - x$ becomes a first-order linear differential equation in the unknown x,

$$x' = -(a + b)x + bm.$$

We recommend that the reader solve the corresponding Cauchy problem with the initial condition $x(0) = x_0$ and analyze the asymptotic behavior of the solution as $t \to +\infty$.

1.3.4 Population Growth of a Single Species

Malthus's Model

Let $p(t)$ be the number of individuals of a single species at time t and let $n(t, p)$ and $m(t, p)$ be the number of individuals born and deceased in the unit time interval $[t, t+1]$. If we agree that the population growth is uniform on short time intervals, then the population change $p(t+h) - p(t)$ in the short time interval $[t, t+h]$ is equal to the difference between the number of births and deaths during that time period, that is $(n(t, p) - m(t, p))h$. Thus

$$\frac{p(t+h) - p(t)}{h} = n(t, p) - m(t, p) \,.$$

Letting $h \to 0$ gives the general form of an equation modeling population growth of a single species

$$p'(t) = n(t, p) - m(t, p) \,. \tag{1.30}$$

This equation yields specific growth laws, if the expression of the terms $n(t, p)$ and $m(t, p)$, or only of their difference $n(t, p) - m(t, p)$, is given. For instance, if we assume that the births and deaths are proportional to the population, that is $n(t, p) = ap(t)$ and $m(t, p) = cp(t)$, then (1.30) becomes *Malthus's*[5] model

$$p' = rp \,, \tag{1.31}$$

where $r = a - c$. Constant a is called the *per capita birth rate*, c is the *per capita death rate*, p' is the *growth rate*, and p'/p is the *per capita growth rate*. The unique solution of equation (1.31) satisfying the initial condition $p(t_0) = p_0$ is the exponential function

$$p(t) = p_0 e^{r(t-t_0)} \,.$$

It is clear that if $a > c$, that is $r > 0$, then the population increases exponentially to infinity as $t \to +\infty$; in contrast, if $a < c$, then the population decreases exponentially to zero tending to disappear; finally if $a = c$, then the population is constant. It is accepted that Malthus's law offers a correct estimation of the population growth on short time intervals. In the long term, the real growth of a population tends to slow down and, under some circumstances, it can even turn into decay. Thus, assuming a constant per capita growth rate is not realistic. A better assumption is to take p'/p be a function of p itself. We shall present two such examples.

The Logistic Equation

If we take

$$\frac{p'}{p} = r\left(1 - \frac{p}{k}\right),$$

---

[5] Thomas Robert Malthus (1766–1834)

then we obtain the *logistic equation*, or the *Verhulst*[6] *model*

$$p' = rp\left(1 - \frac{p}{k}\right), \tag{1.32}$$

where r, k are positive constants. Clearly we may write (1.32) in the form

$$p' = rp - \frac{r}{k}p^2$$

and comparing it to Malthus's model, we can observe the inhibitor effect over the growth rate p' of crowding, mathematically expressed by the quadratic term $-rp^2/k$. We also remark, looking at (1.32), that as long as the population p is under k, the right-hand side of (1.32) is positive, so $p' > 0$ and thus the population increases; in contrast, as long as $p > k$, one has $p' < 0$, so the population decreases. The threshold constant k appears to be a maximum number of individuals that the environment can support and is called the *carrying capacity* of the species. This logistic equation is an example of a model for a self-limiting process, which also arises in other domains of science.

Equation (1.32) has separable variables. We shall solve the Cauchy problem with the initial condition $p(t_0) = p_0$. Separate the variables and use partial fraction decomposition to obtain

$$p'\left(\frac{1}{p-k} - \frac{1}{p}\right) = -r.$$

Integration gives

$$\ln|p-k| - \ln p = -rt + \ln C,$$

so

$$\frac{p-k}{p} = Ce^{-rt}.$$

This leads to the general solution

$$p = \frac{k}{1 - Ce^{-rt}},$$

where C is an arbitrary constant. Under the initial condition, we find the solution

$$p(t) = \frac{k}{1 - \left(1 - \frac{k}{p_0}\right)e^{-r(t-t_0)}}.$$

If we compute $p'(t)$ and then we take the limit of $p(t)$ as $t \to +\infty$, we can see that if $p_0 < k$, then the population increases and approaches k, while if $p_0 > k$, then the population decreases to k.

---

[6] Pierre François Verhulst (1804–1849)

A Cell Population Model

If we take p'/p to be a rational function,

$$\frac{p'}{p} = \frac{a}{1+bp} - c,$$

then we obtain the equation

$$p' = \frac{ap}{1+bp} - cp. \tag{1.33}$$

Here a, b and c are positive constants. This could be a better model for populations with a constant (unchanged) per capita death rate c, for which only the per capita birth rate changes with population. This happens for instance in the case of some cell populations with a fixed lifetime, but whose proliferation in their microenvironment depends on the cell density. Indeed, for equation (1.33), $n(t,p) = ap/(1+bp)$, the ratio $n(t,p)/p = a/(1+bp)$ is almost equal to a for a sparse cell population, and diminishes to zero as p increases. The proportionality constant b is an expression of the cell sensibility to crowding, and is called the *sensibility rate*.

1.3.5 The Gompertz Equation

The *Gompertz[7] equation* is another model for self-limiting growth. In this model the per capita growth rate decreases exponentially with time. The equation is

$$p' = re^{-\alpha t}p, \tag{1.34}$$

where r and α are positive constants. Solving as an equation with separable variables and with the initial condition $p(t_0) = p_0$ yields the growth law

$$p(t) = p_0 \exp\left[\frac{r}{\alpha}\left(e^{-\alpha t_0} - e^{-\alpha t}\right)\right]. \tag{1.35}$$

Notice that the limit of $p(t)$ as $t \to +\infty$ is

$$p_0 \exp\left(\frac{r}{\alpha}e^{-\alpha t_0}\right)$$

and depends on the initial data t_0 and p_0. In contrast, for the Verhulst growth law the limit was k, independent of the values of t_0 and p_0.

Note that if

$$\ln\frac{r}{p_0} = \frac{r}{\alpha}e^{-\alpha t_0}, \tag{1.36}$$

then the function (1.35) is a solution of the autonomous nonlinear equation

$$p' = \alpha p \ln\frac{r}{p}. \tag{1.37}$$

---

[7] Benjamin Gompertz (1779–1865)

Indeed, from (1.35), one has

$$p \ln \frac{r}{p} = p \ln \left(\frac{r}{p_0} \exp \left[-\frac{r}{\alpha} \left(e^{-\alpha t_0} - e^{-\alpha t} \right) \right] \right)$$
$$= p \left(\ln \frac{r}{p_0} - \frac{r}{\alpha} e^{-\alpha t_0} + \frac{r}{\alpha} e^{-\alpha t} \right).$$

Furthermore, in view of (1.36), the first two terms cancel out, so

$$p \ln \frac{r}{p} = p \frac{r}{\alpha} e^{-\alpha t}.$$

Since p is a solution of (1.34), the right-hand side is equal to p'/α, and thus p solves (1.37).

The Gompertz equation has been used as a mathematical model for the growth of animals and the growth of cancer tumors.

Chapter 2 Linear Differential Systems

2.1 Preliminaries

All mathematical models presented so far have been expressed by a single differential equation involving a single time-dependent variable representing, in each case, a quantity of radioactive substance, the temperature of an object, or a population density. However, many evolutionary processes in nature and society cannot be described by only one variable and thus we need to simultaneously consider several differential equations with several dependent variables. We expect that if n dependent variables are considered, then there will be n differential equations connecting them in the same way as algebraic systems do. A set of n first-order differential equations in n dependent variables is called an *n-dimensional differential system*. For $n = 2$, the systems are said to be *two-dimensional* or *planar*.

The normal form of an n-dimensional differential system is

$$\begin{cases} x_1' = f_1(t, x_1, x_2, \ldots, x_n) \\ x_2' = f_2(t, x_1, x_2, \ldots, x_n) \\ \ldots \\ x_n' = f_n(t, x_1, x_2, \ldots, x_n) \,. \end{cases} \quad (2.1)$$

Here $f_1, f_2, \ldots, f_n$ are given functions of $n+1$ real variables, and $x_1, x_2, \ldots, x_n$ are the unknown functions of variable t.

A *solution* of the system (2.1) consists of a set $(x_1, x_2, \ldots, x_n)$ of n functions from $C^1(J)$, where J is some interval of real numbers that satisfy the equations for all $t \in J$, i.e.

$$x_i'(t) = f_i(t, x_1(t), x_2(t), \ldots, x_n(t)) \,,$$

for all $t \in J$ and $i = 1, 2, \ldots, n$.

By the *Cauchy problem* associated with the system (2.1) we mean the problem of finding the solution $(x_1, x_2, \ldots, x_n)$ of the system that satisfies the initial conditions

$$x_1(t_0) = x_1^0, \, x_2(t_0) = x_2^0, \, \ldots, \, x_n(t_0) = x_n^0 \,. \quad (2.2)$$

Here $t_0, x_1^0, x_2^0, \ldots, x_n^0$ are *initial data*, t_0 is a given point of the interval J called the *initial point*, and $x_1^0, x_2^0, \ldots, x_n^0$ are n given numbers called the *initial values*.

The differential systems, just like the differential equations, may be linear or nonlinear. A differential system is said to be *linear* if each component differential equation is linear in the dependent variables. Thus, the n-dimensional differential system (2.1) is linear if it has the form

$$\begin{cases} x_1' = a_{11}(t)x_1 + a_{12}(t)x_2 + \ldots + a_{1n}(t)x_n + b_1(t) \\ x_2' = a_{21}(t)x_1 + a_{22}(t)x_2 + \ldots + a_{2n}(t)x_n + b_2(t) \\ \ldots \\ x_n' = a_{n1}(t)x_1 + a_{n2}(t)x_2 + \ldots + a_{nn}(t)x_n + b_n(t) . \end{cases} \quad (2.3)$$

Here the coefficients a_{ij} and the nonhomogeneous terms b_i ($i, j = 1, 2, \ldots, n$) are functions defined on a given interval J. If the nonhomogeneous terms b_i ($i = 1, 2, \ldots, n$) are zero, then the system is homogeneous.

If, in particular, the coefficients $a_{ij}(i, j = 1, 2, \ldots, n)$ are constant, then we say that the system has *constant coefficients*. Hence an n-dimensional differential system with constant coefficients has the form

$$\begin{cases} x_1' = a_{11}x_1 + a_{12}x_2 + \ldots + a_{1n}x_n + b_1(t) \\ x_2' = a_{21}x_1 + a_{22}x_2 + \ldots + a_{2n}x_n + b_2(t) \\ \ldots \\ x_n' = a_{n1}x_1 + a_{n2}x_2 + \ldots + a_{nn}x_n + b_n(t) . \end{cases} \quad (2.4)$$

For two-dimensional systems, instead of x_1, x_2 and f_1, f_2, the simplest notations x, y and f, g are used in many cases. Thus a two-dimensional differential system is of the form

$$\begin{cases} x' = f(t, x, y) \\ y' = g(t, x, y) , \end{cases}$$

and the corresponding initial conditions in the Cauchy problem can be written as

$$x(t_0) = x_0, \quad y(t_0) = y_0 .$$

Additionally, a two-dimensional linear system takes the form

$$\begin{cases} x' = a_{11}(t)x + a_{12}(t)y + b_1(t) \\ y' = a_{21}(t)x + a_{22}(t)y + b_2(t) , \end{cases} \quad (2.5)$$

and a two-dimensional linear system with constant coefficients can be written as

$$\begin{cases} x' = a_{11}x + a_{12}y + b_1(t) \\ y' = a_{21}x + a_{22}y + b_2(t) . \end{cases} \quad (2.6)$$

2.2 Mathematical Modeling with Linear Differential Systems

A Mathematical Model of an Arms Race

Suppose that $x(t)$ and $y(t)$ are the stocks of arms at time t of two military blocs. Assume that each of the two blocs tries: (i) to increase its arms stock proportional to the armament level of the other bloc; (ii) to decrease the armament level proportional to the corresponding economic burden; and (iii) to take into account the grievances and suspicions of the other bloc. Mathematically, the model can be described by the linear differential system

$$\begin{cases} x' = -a_1 x + b_1 y + g_1 \\ y' = b_2 x - a_2 y + g_2 \, . \end{cases}$$

In this system, x' and y' are the change rates of the armament stocks, and the coefficients are nonnegative; b_1, b_2 stand for the intensities of arms accumulation; a_1, a_2 depend on the economic burden; and g_1, g_2 simulate grievances and suspicions.

This simple model proposed by Richardson[1] is a first approximation of a more complex political and economical process. However, it can give mathematical explanation to some essential aspects of the arms race. Further generalizations have been given in order to take into account additional parameters and conditioning.

A Biological Bicompartmental Model

Compartmental models arise in many areas of science and technology. The compartments may be reservoirs, chemical reactors, organs, the blood system, groups of people, etc.

Assume a drug is administered in a system of two compartments C_1 and C_2 separated by a membrane. The drug can penetrate the membrane in both directions and can also leave compartment C_2. We could for example picture the blood system and an organ. Let v_1, v_2 be the volumes (ml) of the two compartments and let A be the area (dm^2) of the membrane. Denote by $x_1(t)$ and $x_2(t)$ the concentration of the drug (mg per volume) at time t in each of the compartments respectively. Assume that the drug flows from C_1 into C_2 at a rate of q_1 (ml per area per time), and from C_2 into C_1, at the rate q_2. Then $v_1 x_1'$ is the change rate of the drug quantity in the first compartment, and $q_1 A x_1, q_2 A x_2$ are the amounts of drug that at time t penetrate the membrane from C_1 into C_2, and from C_2 into C_1 respectively. Thus the drug balance equation for the first compartment is

$$v_1 x_1' = -q_1 A x_1 + q_2 A x_2 \, .$$

For the second compartment, we have

$$v_2 x_2' = q_1 A x_1 - q_2 A x_2 - k v_2 x_2 \, ,$$

where k (mg per volume per time) is the rate of elimination of the drug from the second compartment. Hence we have arrived at the linear differential system

$$\begin{cases} x_1' = -\frac{q_1 A}{v_1} x_1 + \frac{q_2 A}{v_1} x_2 \\ x_2' = \frac{q_1 A}{v_2} x_1 - \left(\frac{q_2 A}{v_2} + k \right) x_2 \, . \end{cases}$$

---

[1] Lewis Richardson (1881–1953)

Obviously, the model can be generalized to an arbitrary number n, $n \geq 2$, of compartments leading to n-dimensional linear differential systems.

In both examples, the coefficients were assumed to be constant. However, in many cases these coefficients change over time, making the differential systems nonautonomous.

Linearization of Nonlinear Systems

The use of linear differential systems suffices in many applications, but most real phenomena require nonlinear differential systems. Unlike linear systems for which there is a complete theory and high performance numerical software, the nonlinear systems are much more complicated and for some theoretical or practical purposes we use the so-called *linearization* technique. We shall present this technique for the autonomous two-dimensional nonlinear system

$$\begin{cases} x' = f(x, y) \\ y' = g(x, y) . \end{cases} \quad (2.7)$$

We shall linearize the system about a critical point (x_0, y_0), that is a point with

$$f(x_0, y_0) = g(x_0, y_0) = 0 .$$

To this aim, we decompose f and g into a sum of a linear and a nonlinear parts using, for example, Taylor's formula

$$f(x, y) = f_x(x_0, y_0)(x - x_0) + f_y(x_0, y_0)(y - y_0)$$
$$\quad + \text{ higher-order terms in } x - x_0 \text{ and } y - y_0 ,$$
$$g(x, y) = g_x(x_0, y_0)(x - x_0) + g_y(x_0, y_0)(y - y_0)$$
$$\quad + \text{ higher-order terms in } x - x_0 \text{ and } y - y_0 .$$

Then instead of f and g we take only the first-order terms to obtain the nonhomogeneous linear differential system

$$\begin{cases} x' = f_x(x_0, y_0)(x - x_0) + f_y(x_0, y_0)(y - y_0) \\ y' = g_x(x_0, y_0)(x - x_0) + g_y(x_0, y_0)(y - y_0) . \end{cases} \quad (2.8)$$

Here we can change the variables by considering the deviations u, v of x, y from x_0, y_0,

$$u = x - x_0, \quad v = y - y_0 ,$$

and rewrite (2.8) in the form of a homogeneous linear differential system with constant coefficients

$$\begin{cases} u' = f_x(x_0, y_0)u + f_y(x_0, y_0)v \\ v' = g_x(x_0, y_0)u + g_y(x_0, y_0)v , \end{cases}$$

called the *linearization* of (2.7) around (x_0, y_0).

As with Taylor's formula, where the linear part gives a good approximation of the function in a neighborhood of the decomposition point, the linearized system can be seen as a good local approximation of the nonlinear system. We will see later that some of the properties of the associated linearized system can give information about the original nonlinear system.

2.3 Matrix Notation for Systems

The system (2.1) can be written as a single vector equation

$$u' = F(t, u),$$

where u and F are vector-valued functions with n scalar components, or more exactly column vectors, namely

$$u = \begin{bmatrix} x_1 \\ x_2 \\ \vdots \\ x_n \end{bmatrix}, \quad F(t, u) = \begin{bmatrix} f_1(t, x_1, x_2, \ldots, x_n) \\ f_2(t, x_1, x_2, \ldots, x_n) \\ \vdots \\ f_n(t, x_1, x_2, \ldots, x_n) \end{bmatrix}.$$

Additionally, the initial conditions (2.2) can be put in the vector form

$$u(t_0) = u_0,$$

where

$$u_0 = \begin{bmatrix} x_1^0 \\ x_2^0 \\ \vdots \\ x_n^0 \end{bmatrix}.$$

Similarly, the linear system (2.3) has the vector-matrix form

$$u' = A(t)u + b(t), \qquad (2.9)$$

where $A(t)$ is a square matrix and $b(t)$ is a column vector,

$$A(t) = \begin{bmatrix} a_{11}(t) & \cdots & a_{1n}(t) \\ \vdots & & \vdots \\ a_{n1}(t) & \cdots & a_{nn}(t) \end{bmatrix}, \quad b(t) = \begin{bmatrix} b_1(t) \\ \vdots \\ b_n(t) \end{bmatrix}.$$

In particular, the constant coefficients linear system (2.4) is compactly written as

$$u' = Au + b(t),$$

where now $A = [a_{ij}]_{i,j=1,2,\ldots,n}$ is a constant matrix.

2.4 Superposition Principle for Linear Systems

Let us start by observing the identical form of the linear differential system (2.9) and the linear differential equation (1.13). Thus we expect that the conclusions obtained for the linear equations can be extended to linear systems. First, we immediately can see that Theorem 1.6 can be generalized to linear differential systems. Indeed, if we associate (2.9) with the homogeneous system

$$u' = A(t)u, \qquad (2.10)$$

then using the same arguments as for Theorem 1.6, we can prove the *superposition principle* for linear differential systems.

Theorem 2.1. *The general solution u of the linear differential system (2.9) can be represented in the form*

$$u = u_h + u_p,$$

where u_h is the general solution of the homogeneous system (2.10) and u_p is a particular solution of (2.9).

Example 2.2. Verify that the pair $(2t, -t^2)$ is a solution of the system

$$\begin{cases} x' = \frac{2}{t}x + \frac{1}{t^2}y - 1 \\ y' = -2x + 2t, \end{cases} \qquad (2.11)$$

then find its general solution if it is known that the general solution of the homogeneous system

$$\begin{cases} x' = \frac{2}{t}x + \frac{1}{t^2}y \\ y' = -2x \end{cases}$$

is

$$\begin{cases} x = C_1 t + C_2 \\ y = -C_1 t^2 - 2C_2 t - t^2. \end{cases} \qquad (2.12)$$

Solution. According to the superposition principle, the general solution of (2.11) is

$$\begin{cases} x = C_1 t + C_2 + 2t \\ y = -C_1 t^2 - 2C_2 t - t^2, \end{cases} \qquad (2.13)$$

where C_1 and C_2 are arbitrary constants. Note that (2.13) can be put in the vector form

$$u = C_1 \begin{bmatrix} t \\ -t^2 \end{bmatrix} + C_2 \begin{bmatrix} 1 \\ -2t \end{bmatrix} + \begin{bmatrix} 2t \\ -t^2 \end{bmatrix} = \begin{bmatrix} C_1 t + C_2 + 2t \\ -C_1 t^2 - 2C_2 t - t^2 \end{bmatrix}.$$

2.5 Linear Differential Systems with Constant Coefficients

According to the superposition principle, to solve linear systems we need to follow three steps:
- Find the general solution u_h of the associated homogeneous system;
- Find a particular solution u_p of the nonhomogeneous system;
- Write down the general solution as $u = u_h + u_p$.

In what follows, these steps will be completed for systems with constant coefficients. Therefore, we deal with systems of the form

$$u' = Au + b(t), \quad t \in J, \tag{2.14}$$

where A is a square matrix of order n and b is a continuous vector-valued function on J. Additionally, we shall consider the Cauchy problem with the initial condition

$$u(t_0) = u_0, \tag{2.15}$$

where $t_0 \in J$.

2.5.1 The General Solution

Integrating Factor

We proceed formally as in the case of the equation (1.13). We multiply in (2.14) by e^{-tA} (considered now as a matrix), to obtain

$$e^{-tA}u' - e^{-tA}Au = e^{-tA}b(t).$$

Observe that the left-hand side is the total derivative of the product $e^{-tA}u$, which gives

$$(e^{-tA}u)' = e^{-tA}b(t).$$

Then by integration from a point $t_0 \in J$ to an arbitrary t, we obtain

$$e^{-tA}u = \int_{t_0}^{t} e^{-sA}b(s)\,ds + C,$$

where $C = [C_1, C_2, \ldots, C_n]^T$ is an arbitrary column vector. Here, and in what follows, the symbol 'T' is used to denote the transpose of a row vector. Next, multiply by e^{tA} to obtain the general solution

$$u = e^{tA}C + \int_{t_0}^{t} e^{(t-s)A}b(s)\,ds.$$

The constant column vector C is determined by the initial condition

$$u(t_0) = u_0,$$

so that the unique solution of the corresponding Cauchy problem is

$$u = e^{(t-t_0)A} u_0 + \int_{t_0}^{t} e^{(t-s)A} b(s)\, ds.$$

Matrix Exponential

Of course, a rigorous justification should accompany the above formal reasoning. First we must give a sense of the integral and derivative of a matrix-valued function. This is an easy task, since in a natural way these are obtained via term-by-term integration and differentiation of the matrix entries. The result of such an operation is also a matrix of the same type. Next, we have to give a meaning to the symbol e^M, where M is a square matrix of real numbers. For this, it is enough to recall the definition of the ordinary exponential number e^m, where $m \in \mathbb{R}$, as a series of real numbers,

$$e^m = 1 + \frac{1}{1!}m + \frac{1}{2!}m^2 + \ldots + \frac{1}{k!}m^k + \ldots.$$

Now it is easy to guess that the *matrix exponential* e^M is defined similarly by a series of matrices

$$e^M = I + \frac{1}{1!}M + \frac{1}{2!}M^2 + \ldots + \frac{1}{k!}M^k + \ldots, \tag{2.16}$$

where I is the identity matrix of the same order as M.

Let $\|\cdot\|$ denote the Euclidean norm of a vector from $\mathbb{R}^n$ and also the Euclidean norm of a square matrix $M = [a_{ij}]$ of order n, i.e.

$$\|M\| := \left(\sum_{i,j=1}^{n} a_{ij}^2 \right)^{1/2}.$$

Then, using the Cauchy–Schwarz inequality, it is easy to check the following inequalities:

$$\|Mx\| \le \|M\|\, \|x\|,$$
$$\|MN\| \le \|M\|\, \|N\|,$$

for every column vector $x \in \mathbb{R}^n$ and square matrices M, N of order n. Now we are ready to prove the convergence of the series of matrices (2.16) and some basic properties of the matrix exponential.

Proposition 2.3. *Let M and N be square matrices of order n and O be the null square matrix of order n. Then the following statements hold:*
(1) *The series (2.16) is norm convergent and $\|e^M\| \le e^{\|M\|}$.*
(2) *$e^O = I$; if $MN = NM$, then $e^{M+N} = e^M e^N$; e^M is invertible and $(e^M)^{-1} = e^{-M}$.*
(3) *$(e^{tM})' = Me^{tM}$ for all $t \in \mathbb{R}$.*

Proof. (1) The conclusion follows from the comparison of the series of matrices with a convergent numerical series, based on the inequality

$$\left\| \sum_{k=m}^{m+p} \frac{1}{k!} M^k \right\| \le \sum_{k=m}^{m+p} \frac{1}{k!} \|M\|^k .$$

(2) The first property follows directly from the definition of the matrix exponential. Assume that $MN = NM$. Then

$$(M+N)^k = \sum_{j=0}^{k} \binom{k}{j} M^j N^{k-j} ,$$

and

$$e^{M+N} = \sum_{k=0}^{\infty} \frac{1}{k!} (M+N)^k = \sum_{k=0}^{\infty} \frac{1}{k!} \sum_{j=0}^{k} \binom{k}{j} M^j N^{k-j}$$

$$= \sum_{k=0}^{\infty} \frac{1}{k!} \sum_{j=0}^{k} \frac{k!}{j!(k-j)!} M^j N^{k-j}$$

$$= \sum_{k=0}^{\infty} \sum_{i+j=k} \left(\frac{1}{j!} M^j \right) \left(\frac{1}{i!} N^i \right) = e^M e^N .$$

Next, observe that for every $t, s \in \mathbb{R}$, the matrices e^{tM}, e^{sM} commute (first check the commutativity of their partial sums). Then

$$I = e^0 = e^{M+(-M)} = e^M e^{-M} ,$$

which yields $(e^M)^{-1} = e^{-M}$.

(3) We have

$$(e^{tM})' = \lim_{h \to 0} \frac{1}{h} (e^{(t+h)M} - e^{tM}) = \lim_{h \to 0} \frac{1}{h} (e^{hM} - I) e^{tM} .$$

Since

$$e^{hM} - I = \frac{h}{1!} M + \frac{h^2}{2!} M^2 + \ldots ,$$

the limit of $h^{-1}(e^{hM} - I)$ as $h \to 0$ is M. Hence $(e^{tM})' = M e^{tM}$. □

In addition, note that the product rule for derivatives of real-valued functions remains valid for matrix-valued functions.

All the above considerations about matrix-valued functions justify the formal reasoning from the beginning of this section and the conclusions of the next theorem.

Theorem 2.4. (a) (General solution) *For any square matrix A of order n, any continuous function b from J to $\mathbb{R}^n$, and any $t_0 \in J$, the general solution of the n-dimensional linear system (2.14) can be represented in the form*

$$u = e^{tA} C + \int_{t_0}^{t} e^{(t-s)A} b(s) \, ds , \tag{2.17}$$

where C is an arbitrary column vector in $\mathbb{R}^n$.

(b) (Cauchy problem) *The unique solution of the Cauchy problem (2.14)–(2.15) is*

$$u = e^{(t-t_0)A} u_0 + \int_{t_0}^{t} e^{(t-s)A} b(s) \, ds \, . \tag{2.18}$$

2.5.2 Structure of Solution Set for Homogeneous Linear Systems

From formula (2.17), we observe that the general solution of the homogeneous system associated with (2.14),

$$u' = Au \, , \tag{2.19}$$

is

$$u = e^{tA} C \, ,$$

where $C = [C_1, C_2, \ldots, C_n]^T$ is an arbitrary column vector in $\mathbb{R}^n$. As a consequence, if we choose $C = [1, 0, \ldots, 0]^T$, then a particular solution u_1 of the homogeneous system is obtained, which is exactly the first column of the matrix e^{tA}. Similarly, if we choose $C = [0, 1, 0, \ldots, 0]$, then a second particular solution u_2 of the homogeneous system is obtained, namely the second column of the matrix e^{tA}, and so on until u_n. Since the matrix e^{tA} is nonsingular, the solutions $u_1, u_2, \ldots, u_n$ are linearly independent. In addition, from the representation

$$e^{tA} \begin{bmatrix} C_1 \\ C_2 \\ \vdots \\ C_n \end{bmatrix} = C_1 e^{tA} \begin{bmatrix} 1 \\ 0 \\ \vdots \\ 0 \end{bmatrix} + C_2 e^{tA} \begin{bmatrix} 0 \\ 1 \\ \vdots \\ 0 \end{bmatrix} + \ldots + C_n e^{tA} \begin{bmatrix} 0 \\ 0 \\ \vdots \\ 1 \end{bmatrix} ,$$

it follows that any other solution u_h of the homogeneous system is a linear combination of the solutions $u_1, u_2, \ldots, u_n$, that is

$$u_h = C_1 u_1 + C_2 u_2 + \ldots + C_n u_n \, . \tag{2.20}$$

Thus, the set $u_1, u_2, \ldots, u_n$ generates the space of all the solutions of the homogeneous system (2.19). Therefore we may conclude:

Theorem 2.5. *The solution set of a homogeneous linear n-dimensional differential system with constant coefficients is a linear space of dimension n.*

We shall see later that the same is true for linear systems with coefficients that are not necessarily constant.

In addition, we observe that the vector-valued function

$$u_p := \int_{t_0}^{t} e^{(t-s)A} b(s) \, ds$$

is a particular solution of the nonhomogeneous system (2.14).

Example 2.6. Consider the linear system
$$\begin{cases} x' = x - 2y + t \\ y' = x - y. \end{cases} \quad (2.21)$$

Verify that the pair $(t+1, t)$ is a solution of the system, and that the pairs $(\cos t + \sin t, \sin t)$ and $(\cos t - \sin t, \cos t)$ are two solutions of the associated homogeneous system
$$\begin{cases} x' = x - 2y \\ y' = x - y. \end{cases}$$
Then find the general solution of (2.21).

Solution. The general solution of the homogeneous system is
$$\begin{cases} x = C_1 (\cos t + \sin t) + C_2 (\cos t - \sin t) \\ y = C_1 \sin t + C_2 \cos t, \end{cases}$$
and the general solution of the nonhomogeneous system is
$$\begin{cases} x = C_1 (\cos t + \sin t) + C_2 (\cos t - \sin t) + t + 1 \\ y = C_1 \sin t + C_2 \cos t + t. \end{cases}$$
These solutions can be put in the vector form
$$u_h = C_1 \begin{bmatrix} \cos t + \sin t \\ \sin t \end{bmatrix} + C_2 \begin{bmatrix} \cos t - \sin t \\ \cos t \end{bmatrix},$$
and
$$u = C_1 \begin{bmatrix} \cos t + \sin t \\ \sin t \end{bmatrix} + C_2 \begin{bmatrix} \cos t - \sin t \\ \cos t \end{bmatrix} + \begin{bmatrix} t+1 \\ t \end{bmatrix}$$
respectively.

2.5.3 The Concept of Fundamental Matrix

It is clear that in the representation formula (2.20) we may consider another basis $u_1, u_2, \ldots, u_n$ different from the one supplied by the columns of the matrix e^{tA}, and if we denote by $U(t)$ the matrix having the columns $u_1(t), u_2(t), \ldots, u_n(t)$, then the general solution of the homogeneous linear system (2.19) can be written as
$$u = U(t)C,$$
where $C = [C_1, C_2, \ldots, C_n]^T$ is an arbitrary column vector. Such a matrix whose columns are linearly independent solutions of the homogeneous linear system is called a *fundamental* or *transition matrix* of system (2.14). Thus, each basis of the linear space of all the solutions of the homogeneous system yields a fundamental matrix of the system. In particular, the matrix e^{tA} is a fundamental matrix of (2.14).

2.5 Linear Differential Systems with Constant Coefficients

The next proposition gives a characterization of the fundamental matrices and the relationship between different fundamental matrices of the same linear system.

Proposition 2.7. (a) *A square matrix $U(t)$, is a fundamental matrix of the linear system (2.14) if and only if it is nonsingular and*

$$U'(t) = AU(t), \qquad (2.22)$$

for every $t \in J$.
(b) *If $U(t)$ and $V(t)$ are two fundamental matrices of (2.14), then*

$$U(t)U(t_0)^{-1} = V(t)V(t_0)^{-1}, \qquad (2.23)$$

for every $t, t_0 \in J$.

Proof. (a) Assume that $U(t)$ is a fundamental matrix and let $u_1, u_2, \ldots, u_n$ be its columns. From the definition of a fundamental matrix, $u_1, u_2, \ldots, u_n$ are linearly independent solutions of the homogeneous system (2.19). Assume by contradiction that there is a point $t_0 \in J$ for which the matrix $U(t_0)$ is singular. Then the columns of this matrix are linearly dependent, that is there exist real numbers $\alpha_1, \alpha_2, \ldots, \alpha_n$ that are not all zero, such that

$$\alpha_1 u_1(t_0) + \alpha_2 u_2(t_0) + \ldots + \alpha_n u_n(t_0) = 0.$$

Then the function

$$u := \alpha_1 u_1 + \alpha_2 u_2 + \ldots + \alpha_n u_n$$

is a solution of the homogeneous system and $u(t_0) = 0$. The zero function has also these properties. Then, the uniqueness of the solution of the Cauchy problem implies $u = 0$, i.e.

$$\alpha_1 u_1 + \alpha_2 u_2 + \ldots + \alpha_n u_n = 0,$$

which contradicts the linear independence of $u_1, u_2, \ldots, u_n$. Hence $U(t)$ is nonsingular for every t. As regards (2.22), it is enough to observe that this equality read out column-by-column is equivalent to

$$u_1' = Au_1, \quad u_2' = Au_2, \quad \ldots, \quad u_n' = Au_n.$$

(b) Since the columns of the matrices are the basis of the same linear space, there is a column vector C such that

$$U(t) = V(t)C$$

for all $t \in J$. This, for $t = t_0$, gives $C = V(t_0)^{-1}U(t_0)$, where $U(t) = V(t)V(t_0)^{-1}U(t_0)$ and finally (2.23). □

Example 2.8. Find the fundamental matrix $V(t) := e^{tA}$ of the system given in Example 2.6.

Solution. In this case, the coefficient matrix is

$$A = \begin{bmatrix} 1 & -2 \\ 1 & -1 \end{bmatrix},$$

and a fundamental matrix is

$$U(t) = \begin{bmatrix} \cos t + \sin t & \cos t - \sin t \\ \sin t & \cos t \end{bmatrix}.$$

Since $V(0) = e^0 = I = V(0)^{-1}$, (2.23) yields

$$V(t) = U(t)U(0)^{-1}.$$

After some computation we obtain

$$V(t) = \begin{bmatrix} \cos t + \sin t & -2\sin t \\ \sin t & \cos t - \sin t \end{bmatrix}.$$

Solution Representation with Fundamental Matrix

Using the relationship (2.23) we can obtain the analogue of the representation formulas (2.17) and (2.18) in terms of any other fundamental matrix.

Theorem 2.9. *Let $U(t)$ be a fundamental matrix of the system (2.14).*
(a) *(General solution) The general solution of (2.14) is*

$$u = U(t)C + \int_{t_0}^{t} U(t)U(s)^{-1}b(s)\,ds. \tag{2.24}$$

(b) *(Cauchy problem) The unique solution of the Cauchy problem (2.14)-(2.15) is*

$$u = U(t)U(t_0)^{-1}u_0 + \int_{t_0}^{t} U(t)U(s)^{-1}b(s)\,ds. \tag{2.25}$$

Proof. In view of (2.23), we have

$$U(t)U(s)^{-1} = e^{tA}e^{-sA} = e^{(t-s)A}$$

and similarly

$$U(t)U(t_0)^{-1} = e^{(t-t_0)A}.$$

Substitution of $e^{(t-t_0)A}$ and $e^{(t-s)A}$ into (2.17) and (2.18) yields (2.24) and (2.25). □

The computation of the elements of the fundamental matrix e^{tA} from the series (2.16) is almost impossible. Thus a natural question of practical interest is how can we obtain a fundamental matrix of a given system? Our aim in the next subsection is to show that such a fundamental matrix can be determined for any linear system with constant coefficients.

2.5.4 Method of Eigenvalues and Eigenvectors

To begin with we shall only consider two-dimensional systems

$$\begin{cases} x' = a_{11}x + a_{12}y + b_1(t) \\ y' = a_{21}x + a_{22}y + b_2(t) \,, \end{cases} \tag{2.26}$$

where the coefficients a_{ij} ($i, j = 1, 2$) are constant and $b_1, b_2 \in C(J)$. In the vector-matrix form the system reads as

$$u' = Au + b(t)\,, \tag{2.27}$$

where

$$u = \begin{bmatrix} x \\ y \end{bmatrix}, \quad A = \begin{bmatrix} a_{11} & a_{12} \\ a_{21} & a_{22} \end{bmatrix}, \quad b(t) = \begin{bmatrix} b_1(t) \\ b_2(t) \end{bmatrix}.$$

We already know that the general solution has the representation

$$u = u_h + u_p\,,$$

where u_h is the general solution of the associated homogeneous system and u_p is a particular solution of the nonhomogeneous system. Explicitly, the general solution of the system is

$$u = e^{tA}C + \int_{t_0}^{t} e^{(t-s)A} b(s)\, ds\,,$$

which depends on an arbitrary constant vector $C = [C_1, C_2]^T$. Additionally, the unique solution of the Cauchy problem with the initial condition $u(t_0) = u_0$ is

$$u = e^{(t-t_0)A} u_0 + \int_{t_0}^{t} e^{(t-s)A} b(s)\, ds\,. \tag{2.28}$$

We also know that if instead of the fundamental matrix e^{tA} one considers another fundamental matrix $U(t)$, then the general solution and the solution of the Cauchy problem are given by the formulas (2.24) and (2.25). The question that has been asked before is how would we find a fundamental matrix and, consequently, how would we effectively solve the system?

If we compare the homogeneous system

$$u' = Au \tag{2.29}$$

to the scalar equation $x' = ax$ whose solutions are exponential functions, then we would expect exponential-type solutions for the homogeneous system as well. Therefore, we attempt to find a solution of the form

$$u = e^{rt} v\,,$$

where r is a constant and $v = [v_1, v_2]^T$ is a nonzero column vector. The number r and the vector v will be determined by the condition that u satisfies the homogeneous system. Substituting into (2.29) gives

$$re^{rt}v = e^{rt}Av.$$

Simplification by e^{rt} then yields

$$(A - rI)v = 0. \tag{2.30}$$

Hence the number r should be an *eigenvalue* of the matrix A, and v an *eigenvector* associated with the eigenvalue r. The equation (2.30) is a homogeneous linear algebraic system. It has nonzero solutions if and only if the determinant of the coefficient matrix is zero, i.e.

$$\det(A - rI) = 0. \tag{2.31}$$

The equation (2.31) is called the *characteristic equation* of the differential system. Explicitly, the characteristic equation is

$$\begin{vmatrix} a_{11} - r & a_{12} \\ a_{21} & a_{22} - r \end{vmatrix} = 0,$$

which is the quadratic equation

$$r^2 - (a_{11} + a_{22})r + a_{11}a_{22} - a_{12}a_{21} = 0,$$

or equivalently

$$r^2 - (\operatorname{tr} A)r + \det A = 0.$$

Here $\operatorname{tr} A$ is the *trace* of A, defined as the sum of the diagonal elements of A. The roots of the characteristic equation may be real and unequal, real and equal, or complex conjugates. We shall discuss these three cases separately.

Distinct Real Eigenvalues

Assume that the eigenvalues of the matrix A, that is the roots r_1 and r_2 of the characteristic equation, are real and unequal. For each of them, letting r in (2.30) be successively equal with r_1 and r_2, we choose a corresponding eigenvector, v_1 and v_2. Thus we obtain two solutions

$$u_1 = e^{r_1 t} v_1, \quad u_2 = e^{r_2 t} v_2$$

of the homogeneous differential system (2.29). Since $r_1 \neq r_2$, these solutions are linearly independent, so u_1, u_2 is a basis of the solution space, and the matrix $U(t)$, having as columns u_1 and u_2, is a fundamental matrix of the differential system. Therefore, the general solution of (2.29) is

$$u = C_1 e^{r_1 t} v_1 + C_2 e^{r_2 t} v_2,$$

where C_1 and C_2 are two arbitrary constants.

Example 2.10. Solve the homogeneous linear differential system

$$\begin{cases} x' = 4x - 5y \\ y' = x - 2y. \end{cases}$$

Solution. In this case, the characteristic equation of the system

$$r^2 - (\operatorname{tr} A)r + \det A = 0$$

is $r^2 - 2r - 3 = 0$. By the quadratic formula the eigenvalues are $r_1 = -1$ and $r_2 = 3$. For $r = -1$, we write down the algebraic system $(A - rI)v = 0$, that is

$$\begin{cases} (4-r)v_1 - 5v_2 = 0 \\ v_1 + (-2-r)v_2 = 0. \end{cases}$$

As the determinant is zero, the component equations are equivalent, so we may take only one of them, for example the second one:

$$v_1 - v_2 = 0.$$

If we choose $v_1 = 1$, then we obtain $v_2 = 1$. Thus a solution of the differential system is

$$u_1 = e^{-t} \begin{bmatrix} 1 \\ 1 \end{bmatrix}.$$

Similarly, for $r = 3$, we obtain another solution

$$u_2 = e^{3t} \begin{bmatrix} 5 \\ 1 \end{bmatrix}.$$

Consequently, all the solutions of the differential system are the linear combinations of u_1 and u_2, and the general solution is

$$u = C_1 u_1 + C_2 u_2,$$

or explicitly,

$$\begin{cases} x = C_1 e^{-t} + 5C_2 e^{3t} \\ y = C_1 e^{-t} + C_2 e^{3t}, \end{cases}$$

where C_1 and C_2 are two arbitrary constants.

Equal Real Eigenvalues

Assume now that the eigenvalues of A are equal. Then

$$r := r_1 = r_2 = \frac{\operatorname{tr} A}{2}.$$

One solution of the differential system can be obtained as shown previously, of the form

$$u_1 = e^{rt}v,$$

where v is a corresponding eigenvector of A, i.e. a nonzero solution of the homogeneous algebraic system

$$(A - rI)v = 0.$$

In the trivial case $A - rI = 0$, where the differential system is uncoupled, two linearly independent solutions are

$$u_1 = e^{rt}\begin{bmatrix} 1 \\ 0 \end{bmatrix}, \quad u_2 = e^{rt}\begin{bmatrix} 0 \\ 1 \end{bmatrix}.$$

In the nontrivial case $A - rI \neq 0$, a second solution u_2, linearly independent of u_1, is tried in the form

$$u_2 = e^{rt}(tv + w),$$

where v is that from above and w is to be determined. Since

$$u_2' = re^{rt}(tv + w) + e^{rt}v,$$

substitution of u_2 into the system gives

$$re^{rt}(tv + w) + e^{rt}v = e^{rt}A(tv + w).$$

If we divide by e^{rt} and we cancel the equal terms rtv and Atv, we obtain the nonhomogeneous algebraic system

$$(A - rI)w = v.$$

This system is always compatible according to the Kronecker–Capelli theorem, since the rank of its coefficient matrix $A - rI$ is equal to the rank of the augmented matrix $[A - rI \mid V]$ (we leave the proof of this statement as an exercise for the reader). So, a solution w can be found.

Therefore, the general solution of the differential system is the linear combination

$$u = C_1 e^{rt}v + C_2 e^{rt}(tv + w).$$

Example 2.11. Solve the homogeneous differential system

$$\begin{cases} x' = 5x - 3y \\ y' = 3x - y, \end{cases}$$

and find the solution of the Cauchy problem with the initial conditions

$$x(0) = 3, \quad y(0) = 2.$$

Solution. One has $\operatorname{tr} A = 4$ and $\det A = 4$, so the characteristic equation is $r^2 - 4r + 4 = 0$, and has the double root $r = 2$. The first equation of the algebraic system $(A - rI)v = 0$

2.5 Linear Differential Systems with Constant Coefficients

is $3v_1 - 3v_2 = 0$. With the choice $v_1 = 1$, we obtain the eigenvector $v = [1, 1]^T$. Furthermore, as the system $(A - rI)w = v$ has the rank equal to 1, one may consider only one of its equations; let this be the first $3w_1 - 3w_2 = 1$. If we choose for example $w_2 = 1$, then we find $w_1 = 4/3$. Thus we have obtained the solutions

$$u_1 = e^{2t}\begin{bmatrix}1\\1\end{bmatrix}, \quad u_2 = e^{2t}\begin{bmatrix}t + \frac{4}{3}\\t + 1\end{bmatrix}.$$

The general solution of the differential system is

$$u = C_1 u_1 + C_2 u_2,$$

or explicitly

$$\begin{cases} x = C_1 e^{2t} + C_2 \left(t + \frac{4}{3}\right) e^{2t} \\ y = C_1 e^{2t} + C_2 (t + 1) e^{2t}, \end{cases} \tag{2.32}$$

where C_1 and C_2 are two arbitrary constants.

To solve the Cauchy problem, just take $t = 0$ in (2.32). Then

$$C_1 + \frac{4}{3} C_2 = 3, \quad C_1 + C_2 = 2.$$

Solving in C_1 and C_2 gives $C_1 = -1$ and $C_2 = 3$. Therefore the solution of the Cauchy problem is

$$\begin{cases} x = -e^{2t} + (3t + 4)e^{2t} \\ y = -e^{2t} + 3(t + 1)e^{2t}. \end{cases}$$

Complex Eigenvalues

Finally assume that the roots of the characteristic equation are the complex conjugates $\alpha \pm \beta i$. Taking one of them, for instance $r = \alpha + \beta i$, and proceeding as in the case of real unequal eigenvalues, we find a solution, this time as a complex function

$$u = e^{rt}v = e^{(\alpha + i\beta)t}(v_1 + iv_2) = e^{\alpha t}(\cos\beta t + i\sin\beta t)(v_1 + iv_2)$$
$$= e^{\alpha t}[v_1 \cos\beta t - v_2 \sin\beta t + i(v_1 \sin\beta t + v_2 \cos\beta t)],$$

where v_1, v_2 are the column vectors of the real and imaginary parts of v, respectively, that is

$$v = v_1 + iv_2 \quad \text{and} \quad v_1, v_2 \in \mathbb{R}^2.$$

It is easy to see that if u is a complex solution of the system, then the conjugate function $\bar{u}$ is a solution as well. In addition, any linear combination of two solutions, such as with complex coefficients, is also a solution. As a result, the functions

$$u_1 := \operatorname{Re} u = \frac{1}{2}(u + \bar{u}) = e^{\alpha t}(v_1 \cos\beta t - v_2 \sin\beta t),$$
$$u_2 := \operatorname{Im} u = \frac{1}{2i}(u - \bar{u}) = e^{\alpha t}(v_1 \sin\beta t + v_2 \cos\beta t)$$

are real solutions of the system. Therefore, the general solution of the differential system is

$$u = C_1 u_1 + C_2 u_2.$$

Example 2.12. Find the general solution of the system

$$\begin{cases} x' = 5x - 9y \\ y' = 2x - y. \end{cases}$$

Solution. The characteristic equation $r^2 - 4r + 13 = 0$ has the complex roots $2 \pm 3i$. For $r = 2 + 3i$, we look for a nonzero solution of the algebraic system $(A - rI)v = 0$. The first equation of this system is $(3 - 3i)v_1 - 9v_2 = 0$. If we choose $v_1 = 3$ (any other choice $v_1 \neq 0$ is possible), then we find $v_2 = 1 - i$. Hence

$$v = \begin{bmatrix} 3 \\ 1 - i \end{bmatrix} = \begin{bmatrix} 3 \\ 1 \end{bmatrix} + i \begin{bmatrix} 0 \\ -1 \end{bmatrix} = v_1 + iv_2.$$

Thus a complex solution of the system is

$$u = e^{(2+3i)t} v = e^{2t} (\cos 3t + i \sin 3t) \begin{bmatrix} 3 \\ 1 - i \end{bmatrix}$$

$$= e^{2t} \begin{bmatrix} 3 \cos 3t + 3i \sin 3t \\ \cos 3t + \sin 3t + i(\sin 3t - \cos 3t) \end{bmatrix}.$$

Two real solutions are the real and imaginary parts

$$u_1 = \begin{bmatrix} 3e^{2t} \cos 3t \\ e^{2t} (\cos 3t + \sin 3t) \end{bmatrix}, \quad u_2 = \begin{bmatrix} 3e^{2t} \sin 3t \\ e^{2t} (\sin 3t - \cos 3t) \end{bmatrix}.$$

Thus all the real solutions of the system are

$$u = C_1 u_1 + C_2 u_2,$$

or explicitly

$$\begin{cases} x = 3e^{2t} (C_1 \cos 3t + C_2 \sin 3t) \\ y = e^{2t} [C_1 (\cos 3t + \sin 3t) + C_2 (\sin 3t - \cos 3t)], \end{cases}$$

where C_1 and C_2 are two arbitrary real constants.

2.6 Method of Variation of Parameters

Returning to the nonhomogeneous linear system (2.26), or to its vector-matrix formulation (2.27), now that the general solution u_h of the associated homogeneous system

2.6 Method of Variation of Parameters

is obtained, it remains to determine a particular solution u_p of the nonhomogeneous system. Here again we shall extend to systems a method already used for linear differential equations, namely the method of variation of parameters.

Assume that we know two linearly independent solutions

$$u_1 = [x_1, y_1]^T, \quad u_2 = [x_2, y_2]^T$$

of the homogeneous system (2.29), so that the general solution of (2.29) is

$$u_h = C_1 u_1 + C_2 u_2,$$

where C_1, C_2 are arbitrary constants. Then we look for a solution of (2.27) in the form

$$u_p = C_1(t) u_1 + C_2(t) u_2,$$

where the parameters C_1, C_2 have been replaced by t-dependent real-valued functions $C_1(t)$ and $C_2(t)$. These functions will be determined by the requirement that u_p satisfies (2.27). Substitution of u_p into (2.27) yields

$$C_1(t) u_1' + C_2(t) u_2' + C_1'(t) u_1 + C_2'(t) u_2 = A(C_1(t) u_1 + C_2(t) u_2) + b(t),$$

or

$$C_1(t) u_1' + C_2(t) u_2' + C_1'(t) u_1 + C_2'(t) u_2 = C_1(t) A u_1 + C_2(t) A u_2 + b(t).$$

Since u_1 and u_2 are solutions of the homogeneous system, a part of the terms cancel down, giving

$$C_1'(t) u_1 + C_2'(t) u_2 = b(t),$$

or explicitly

$$\begin{cases} C_1'(t) x_1(t) + C_2'(t) x_2(t) = b_1(t) \\ C_1'(t) y_1(t) + C_2'(t) y_2(t) = b_2(t). \end{cases} \quad (2.33)$$

Since any fundamental matrix $U(t)$ is nonsingular for every t, the determinant of the above algebraic system is different from zero, so that $C_1'(t)$ and $C_2'(t)$ are uniquely determined. Finally $C_1(t)$ and $C_2(t)$ are chosen among the primitives of $C_1'(t)$ and $C_2'(t)$, respectively.

Example 2.13. Find a particular solution of the system

$$\begin{cases} x' = 4x - 5y + 1 + 5e^t \\ y' = x - 2y + 1 + e^t. \end{cases}$$

Solution. The associated homogeneous system was solved in Example 2.10. Hence we may take

$$u_1 = \begin{bmatrix} e^{-t} \\ e^{-t} \end{bmatrix}, \quad u_2 = \begin{bmatrix} 5e^{3t} \\ e^{3t} \end{bmatrix}.$$

Then the algebraic system (2.33) becomes

$$\begin{cases} C_1'(t)e^{-t} + 5C_2'(t)e^{3t} = 1 + 5e^t \\ C_1'(t)e^{-t} + C_2'(t)e^{3t} = 1 + e^t . \end{cases}$$

Solving gives

$$C_1'(t) = e^t, \quad C_2'(t) = e^{-2t} .$$

Hence we may choose

$$C_1(t) = e^t, \quad C_2(t) = -\frac{1}{2}e^{-2t} .$$

Thus, a particular solution of the differential system is $u_p = e^t u_1 - 2^{-1} e^{-2t} u_2$, or

$$u_p = \begin{bmatrix} 1 - \frac{5}{2}e^t \\ 1 - \frac{1}{2}e^t \end{bmatrix} .$$

2.7 Higher-Dimensional Linear Systems

The technique based on eigenvalues and eigenvectors can easily be extended to n-dimensional homogeneous linear differential systems with constant coefficients

$$u' = Au , \qquad (2.34)$$

for any $n \geq 2$. Here

$$u = \begin{bmatrix} x_1 \\ x_2 \\ \vdots \\ x_n \end{bmatrix}, \quad A = [a_{ij}]_{i,j=1,2,\ldots,n} .$$

In this case the characteristic equation $\det(A - rI) = 0$ is a polynomial equation of degree n. Consequently, the matrix A has n real or complex not necessary distinct eigenvalues $r_1, r_2, \ldots, r_n$.

If all the eigenvalues are real and distinct, then n linearly independent solutions of the differential system are

$$u_1 = e^{r_1 t} v_1, u_2 = e^{r_2 t} v_2, \ldots, u_n = e^{r_n t} v_n ,$$

where v_i is an eigenvector corresponding to the eigenvalue r_i, i.e. $v_i \neq 0$ and

$$(A - r_i I)v_i = 0, \quad i = 1, 2, \ldots, n .$$

Then the general solution of the differential system is

$$u = C_1 e^{r_1 t} v_1 + C_2 e^{r_2 t} v_2 + \ldots + C_n e^{r_n t} v_n .$$

2.7 Higher-Dimensional Linear Systems — 43

Example 2.14. Solve the system

$$\begin{cases} x' = x + z \\ y' = x + y \\ z' = 2x. \end{cases}$$

Solution. Some computation gives the eigenvalues $r_1 = 1, r_2 = -1$ and $r_3 = 2$. For each of these numbers we look for a nontrivial solution $v = [\alpha, \beta, \gamma]^T$ of the algebraic system $(A - rI)v = 0$, or

$$\begin{cases} (1 - r)\alpha + \gamma = 0 \\ \alpha + (1 - r)\beta = 0 \\ 2\alpha - r\gamma = 0. \end{cases}$$

First, for $r = 1$ we choose a solution $v_1 = [0, 1, 0]^T$; next $r = -1$ gives $v_2 = [2, -1, -4]^T$; and $r = 2$ yields $v_3 = [1, 1, 1]^T$. The general solution is

$$u = C_1 e^t v_1 + C_2 e^{-t} v_2 + C_3 e^{2t} v_3,$$

or

$$\begin{cases} x = -C_2 e^{-t} + C_3 e^{2t} \\ y = C_1 e^t - C_2 e^{-t} + C_3 e^{2t} \\ z = -4C_2 e^{-t} + C_3 e^{2t}, \end{cases}$$

where C_1, C_2 and C_3 are arbitrary constants.

For the case of repeated eigenvalues we use the following result, whose proof, based on the Jordan canonical form of a matrix, is sketched in the next section and can be omitted at first reading.

Theorem 2.15. *For each eigenvalue r (possibly complex) of the matrix A with the order of multiplicity m ($1 \le m \le n$), there exist m linearly independent solutions of the system, of the form*

$$u = e^{rt} Q(t), \qquad (2.35)$$

where $Q(t)$ is a polynomial of degree at most $m - 1$ with coefficients in $\mathbb{C}^n$, i.e.

$$Q(t) = v_0 + tv_1 + \ldots + t^{m-1} v_{m-1}, \quad v_j \in \mathbb{C}^n \ (0 \le j \le m - 1).$$

In the case where the eigenvalue r is a complex number, the solutions (2.35) are complex functions. Then $2m$ real solutions are associated with the pair of complex conjugate eigenvalues r and $\bar{r}$ by taking the real and imaginary parts of the functions (2.35).

Theorem 2.15 yields the following remark, which will be useful in Section 5.3.

Remark 2.16. (a) All solutions of the system (2.34) tend to zero as $t \to +\infty$ if and only if the real parts of the eigenvalues of the matrix A are strictly negative.
(b) All solutions of the system (2.34) are bounded on $\mathbb{R}_+$ if the real parts of the eigenvalues of the matrix A are nonpositive and the eigenvalues having the real part zero are simple.

(c) If at least one of the eigenvalues of the matrix A has a strictly positive real part, then the system has solutions u with $\|u(0)\|$ as small as desired and $\|u(t)\| \to +\infty$ as $t \to +\infty$.

Note that the method of variation of parameters for finding a particular solution of a nonhomogeneous linear system can also be extended to n-dimensional linear systems.

2.8 Use of the Jordan Canonical Form of a Matrix

A fundamental result from linear algebra, Jordan[2] decomposition theorem, asserts that for every square complex matrix A there exists a nonsingular matrix P such that

$$A = P^{-1}JP,$$

where the matrix J, called the *Jordan canonical form* of A has the structure

$$J = \begin{bmatrix} J_1 & & 0 \\ & \ldots & \\ 0 & & J_q \end{bmatrix},$$

with

$$J_k = \begin{bmatrix} r_k & 1 & 0 & \ldots & 0 \\ 0 & r_k & 1 & \ldots & 0 \\ 0 & 0 & r_k & \ldots & 0 \\ \ldots & \ldots & \ldots & \ldots & 1 \\ 0 & 0 & 0 & \ldots & r_k \end{bmatrix}$$

($k = 1, \ldots, q$) having the same element on the diagonal, 1s on the superdiagonal and 0s elsewhere. The matrices J_k are called *Jordan blocks* and their sizes $m_1, \ldots, m_q$ are between 1 and n, with $m_1 + \ldots + m_q = n$. The blocks of size 1 are 1×1 matrices. The proof of Jordan's decomposition theorem is not easy, but it can be found for example in the book by P. R. Halmos, *Finite-Dimensional Vector Spaces*, Springer, New York, 1987.

The diagonal elements of the Jordan blocks are not necessarily different, and it is easy to see that they are the eigenvalues of the matrix J.

The matrices A and J are similar in the following sense:
(1) They have the same characteristic equation, hence the same eigenvalues with the same orders of multiplicity.
Indeed, from
$$A - rI = P^{-1}JP - rP^{-1}P = P^{-1}(J - rI)P \tag{2.36}$$
it follows that the equation $\det(A - rI) = 0$ is equivalent to $\det(J - rI) = 0$.

---

[2] Camille Jordan (1838–1922)

(2) If a vector v is an eigenvector of A, then $w = Pv$ is an eigenvector of J corresponding to the same eigenvalue. Conversely, if w is an eigenvector of J, then $v = P^{-1}w$ is an eigenvector of A.

Indeed, from (2.36), one has that $(A - rI)v = 0$ if and only if $(J - rI)Pv = 0$.

(3) The sum of the sizes of all Jordan blocks that have on the diagonal the same number r (or equivalently, the number of appearances of r on the diagonal of J) is the order of multiplicity of the eigenvalue r of the matrices A and J.

(4) The number of Jordan blocks that have the same number r on the diagonal represents the number of linearly independent eigenvectors of the matrices A and J corresponding to the eigenvalue r.

(5) x solves the differential system $x' = Ax$ if and only if $y = Px$ solves the system $y' = Jy$.

Indeed,
$$y' = Px' = PAx = PP^{-1}JPx = JPx = Jy.$$

In view of the above properties, the proof of Theorem 2.15 reduces to the case where the matrix is a Jordan canonical matrix. In such a case, we are able to do more, namely that for an eigenvalue r of multiplicity m to find explicitly m linearly independent solutions of the system
$$y' = Jy.$$

In fact, to each Jordan block corresponding to the eigenvalue r we shall attach a number of linearly independent solutions equal to the size of that Jordan block.

Let us concentrate our attention on the Jordan block of size m_j of the matrix $J - rI$ and assume that the position of that block is between the $(a + 1)$th and $(a + m_j)$th rows (equivalently, columns). First observe what this Jordan block becomes when the matrix $J - rI$ is taken to a power. If $m_j = 1$, i.e. the block is of size 1, then it remains constant zero. If $m_j \geq 2$, then in $(J - rI)^2$ the superdiagonal of 1s of the initial block moves up one position, in $(J - rI)^3$ it moves one position more, and in $(J - rI)^{m_j-1}$ the entries of the block are zero except the entry in row $a + 1$ and column $a + m_j$, which is equal to one. Finally, the corresponding block of the power $(J - rI)^{m_j}$ is zero. Hence the $(a + m_j)$th column of $(J - rI)^{m_j-1}$ has all elements zero except the element in row $a + 1$, which is 1, while the same column of the matrix $(J - rI)^{m_j}$ is zero. As a result, if we take the vector $v_{j0} = [0, \ldots, 1, \ldots, 0]^T$, where 1 is at position $a + m_j$, then

$$v_{j1} := (J - rI)^{m_j-1} v_{j0} = [0, \ldots, 1, \ldots, 0]^T \neq 0, \tag{2.37}$$

with 1 at the $(a + 1)$th position, and

$$(J - rI)^{m_j} v_{j0} = 0.$$

This shows that v_{j1} is an eigenvector. Next it is easy to verify that if we denote

$$v_{ji} = (J - rI)^{m_j-i} v_{j0}, \quad i = 1, \ldots, m_j,$$

then the following m_j functions are linearly independent solutions of the system $y' = Jy$:

$$e^{rt}v_{j1}, \quad e^{rt}\left(\frac{t}{1!}v_{j1} + v_{j2}\right), \quad r^{rt}\left(\frac{t^2}{2!}v_{j1} + \frac{t}{1!}v_{j2} + v_{j3}\right),$$

$$\ldots, \quad r^{rt}\left(\frac{t^{m_j-1}}{(m_j-1)!}v_{j1} + \ldots + v_{jm_j}\right).$$

Since in virtue of (2.37) the eigenvectors of type v_{j1} corresponding to all Jordan blocks associated with r are linearly independent, we may assert that the m solutions found in this way are linearly independent. Thus Theorem 2.15 is proved.

2.9 Dynamic Aspects of Differential Systems

Most of the mathematical models of real phenomena given by differential equations and systems follow the evolution in time of one, two or several quantities or parameters. This has been the reason in this book for using the notation t for the independent variable. Assigning the meaning of 'time' to the independent variable t introduces from the beginning the dynamic aspect in the theory of differential equations and systems. In addition, the notion of an 'initial condition', used in the formulation of the Cauchy problem, is itself tributary to the category of 'time'. When t means time, the initial point t_0 of an initial condition is often called the *initial time*. We can also use the terminology of *states* for the values of the solutions.

Orbits. Phase Portrait
Let us consider a two-dimensional autonomous system that is not necessarily linear

$$\begin{cases} x' = f(x, y) \\ y' = g(x, y), \end{cases}$$

with the unknown functions x and y of the time variable t, and let us focus our attention on a solution (x, y) of this system. The way that the quantities x and y change with time t and relate to each other at different times can be visualized by the graphical representation, or *time series* plot of $x(t)$ and $y(t)$ as functions of t in the same coordinate plane. Alternatively, we may represent in the xy-coordinate plane, called the *phase plane*, the one-parameter curve of equations $x = x(t), y = y(t)$. Such a curve is called an *orbit*, a *trajectory*, or a *solution curve*. Occasionally, by an arrow, one shows the direction in which the point with coordinates $x(t)$ and $y(t)$ moves along the curve as time increases. The set of all the orbits is called the *phase portrait* of the differential system. Since the representation of all orbits is impossible, a finite number of them is enough to have an idea about it. These representative orbits are obtained as solutions of the Cauchy problems corresponding to a few initial conditions.

Let us note that for each t, the vector $(f(x(t), y(t)), g(x(t), y(t)))$, or equivalently $(x'(t), y'(t))$ is tangent to the orbit at the point with coordinates $x(t)$ and $y(t)$, and its sense is given by the orbit direction. The set of all vectors $\vec{v}(f(x, y), g(x, y))$, where (x, y) belongs to the domain of the functions f and g is called the *vector field* (or *direction field*, or *slope field*) of the differential system. Computer representations of phase portraits often contain the normalized vector field, i.e. vectors of equal size. The sense of these vectors indicates the direction of the orbits.

For some systems, there are also degenerate orbits, which reduce to a single point. These orbits correspond to constant solutions, also called *stationary* or *equilibrium solutions*, *critical points*, or *fixed points*. One can find the equilibria making $x' = y' = 0$, thus solving the system

$$\begin{cases} f(x, y) = 0, \\ g(x, y) = 0. \end{cases}$$

Phase Portrait for Homogeneous Linear Systems with Constant Coefficients

Let us first consider two examples of systems whose orbits are representative geometrical curves given by Cartesian equations.

Example 2.17. Find the orbits of the uncoupled system

$$\begin{cases} x' = x \\ y' = 2y. \end{cases}$$

Solution. The general solution of the system is

$$\begin{cases} x = C_1 e^t \\ y = C_2 e^{2t} \end{cases}$$

and represents the parametrized equations of its orbits. The elimination of t gives the equation $y = C_1^{-2} C_2 x^2$. If we observe that the sign of x is given by the sign of C_1, the sign of y by the sign of C_2, and that x and y tend to zero as $t \to -\infty$, we may conclude that the orbits of the system are semiparabolas of equations

$$y = ax^2, \quad a \in \mathbb{R},$$

which come out of the origin. For $C_1 = C_2 = 0$ we have the null solution as an equilibrium and its corresponding orbit is the origin.

Example 2.18. Find the orbits of the system

$$\begin{cases} x' = 4y \\ y' = -x. \end{cases}$$

Solution. Multiply by x the first equation and by $4y$ the second one, then add the two equations to obtain
$$xx' + 4yy' = 0,$$
or equivalently
$$(x^2 + 4y^2)' = 0.$$
Hence the Cartesian equation of the orbits is
$$x^2 + 4y^2 = C,$$
where C is an arbitrary nonnegative constant. So, the orbits are ellipses with centers at the origin. For $C = 0$, the orbit reduces to the origin of the plane and corresponds to the unique equilibrium of the system. Of course, we may obtain the orbits by solving the system. Then we find the general solution
$$\begin{cases} x = 2C_1 \cos 2t + 2C_2 \sin 2t \\ y = -C_1 \sin 2t + C_2 \cos 2t \end{cases}$$
representing the parametric equations of the orbits. Eliminating t we find their Cartesian equation. Let us focus our attention on a particular orbit, say $x^2 + 4y^2 = 2$. In order to determine its direction it is enough to choose one of its points, for instance $(1, 1/2)$, and then write down the tangent vector to the orbit at this point $\vec{v}(4y, -x) = \vec{v}(2, -1)$. This vector shows the clockwise direction of the orbit.

Coming back to the general case of planar homogeneous linear systems with constant coefficients
$$u' = Au,$$
we start our analysis by noting that the degenerate orbits that reduce to a single point correspond to the stationary solutions and thus to the solutions of the algebraic system
$$Au = 0.$$
If rank $A = 2$, i.e. $\det A \neq 0$, then the system has only the zero solution; if rank $A = 1$, then the set of all the solutions of this system is a linear space of dimension 1, which geometrically in the phase plane corresponds to a line through the origin. The points of this line are the degenerate orbits of the differential system.

Furthermore, the orbit analysis should take into account the three cases depending on the nature of the eigenvalues of the matrix A: real and unequal, real and equal, and complex.

(1) Assume that the eigenvalues r_1 and r_2 are real and unequal. Then the general solution looks like
$$u = C_1 e^{r_1 t} v_1 + C_2 e^{r_2 t} v_2,$$

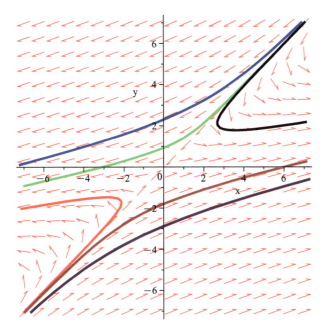

Fig. I.2.1: Phase portrait of the system in Example 2.10.

and thus the parametric equations of the orbits are

$$\begin{cases} x = C_1 e^{r_1 t} v_{11} + C_2 e^{r_2 t} v_{21} \\ y = C_1 e^{r_1 t} v_{12} + C_2 e^{r_2 t} v_{22}, \end{cases}$$

where we have denoted $v_1 = [v_{11}, v_{12}]^T$ and $v_2 = [v_{21}, v_{22}]^T$. In the case $r_1 < 0$ and $r_2 < 0$, one has that $x, y \to 0$ as $t \to +\infty$, which means that the direction of the orbits is towards the origin – the orbit of the zero solution, hence *the orbits approach the origin*. In contrast, if both eigenvalues are positive, i.e. $r_1 > 0$ and $r_2 > 0$, then the direction of the orbits is opposite and *the orbits come out of the origin*. In both cases the origin-orbit of the zero equilibrium is called a *node*. If the eigenvalues have opposite signs, then the origin-orbit of the zero equilibrium is called a *saddle point* and the phase portrait looks like that in Figure I.2.1. In the case where $r_1 \neq 0$ but $r_2 = 0$, the parametric equations of the orbits have the form

$$\begin{cases} x = C_1 e^{r_1 t} v_{11} + C_2 v_{21} \\ y = C_1 e^{r_1 t} v_{12} + C_2 v_{22}, \end{cases}$$

and the elimination of t yields the Cartesian equation of a pencil of parallel lines:

$$\frac{y - C_2 v_{22}}{v_{12}} = \frac{x - C_2 v_{21}}{v_{11}}$$

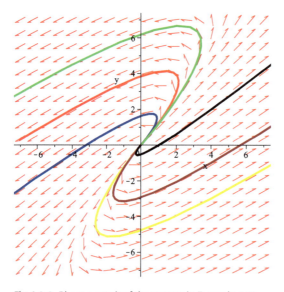

Fig. I.2.2: Phase portrait of the system in Example 2.11.

in the case that $v_{11} \neq 0$ and $v_{12} \neq 0$;

$$x = C_2 v_{21}$$

if $v_{11} = 0$; and

$$y = C_2 v_{22}$$

when $v_{12} = 0$. Therefore, in this case the orbits are semilines.

(2) Assume that the eigenvalues are equal, i.e. $r := r_1 = r_2$. Then the general solution has the form

$$u = C_1 e^{rt} v + C_2 e^{rt}(tv + w),$$

where the column vectors v and w are solutions of the algebraic systems

$$(A - rI)v = 0 \quad \text{and} \quad (A - rI)w = v$$

respectively. If $v = [v_1, v_2]^T$ and $w = [w_1, w_2]^T$, then the parametric equations of the orbits are

$$\begin{cases} x = C_1 e^{rt} v_1 + C_2 e^{rt}(tv_1 + w_1) \\ y = C_1 e^{rt} v_2 + C_2 e^{rt}(tv_2 + w_2). \end{cases}$$

In the case that $r > 0$ (Figure I.2.2), the orbits approach the origin as $t \to -\infty$, while from

$$\frac{y}{x} = \frac{C_1 v_2 + C_2(tv_2 + w_2)}{C_1 v_1 + C_2(tv_1 + w_1)}$$

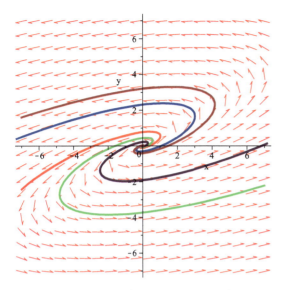

Fig. I.2.3: Phase portrait of the system in Example 2.12.

we see that $y/x \to v_2/v_1$ (when $v_1 \neq 0$) and $x/y \to v_1/v_2$ (when $v_2 \neq 0$) as $t \to +\infty$, that is the orbits asymptotically approach a pencil of parallel lines as time increases. If $r < 0$, the orbit direction is opposite.

(3) Assume that the eigenvalues are complex conjugates, $r = \alpha \pm \beta i$. In this case we know that the solutions read as

$$\begin{cases} x = e^{\alpha t}[C_1(v_{11}\cos\beta t - v_{21}\sin\beta t) + C_2(v_{11}\sin\beta t + v_{21}\cos\beta t)] \\ y = e^{\alpha t}[C_1(v_{12}\cos\beta t - v_{22}\sin\beta t) + C_2(v_{12}\sin\beta t + v_{22}\cos\beta t)]. \end{cases}$$

The terms that contain trigonometric functions are periodic with period $2\pi/\beta$, making it so that the orbits turn round the origin. The factor $e^{\alpha t}$ acts as an amplitude factor. If $\alpha > 0$ (Figure I.2.3) then the orbits are spirals, which go away from the origin as t increases, while if $\alpha < 0$, then the amplitude factor $e^{\alpha t}$ decreases to zero and so the spirals approach the origin as t increases. When $\alpha \neq 0$ we say that the origin is a *spiral point* or a *focus*. A particular interesting case is $\alpha = 0$, when the eigenvalues are purely imaginary. Then the amplitude factor is equal to 1 and the orbits

$$\begin{cases} x = C_1(v_{11}\cos\beta t - v_{21}\sin\beta t) + C_2(v_{11}\sin\beta t + v_{21}\cos\beta t) \\ y = C_1(v_{12}\cos\beta t - v_{22}\sin\beta t) + C_2(v_{12}\sin\beta t + v_{22}\cos\beta t) \end{cases}$$

are closed curves around the origin. Solving this system in $\sin\beta t$ and $\cos\beta t$ and using the formula $\sin^2\beta t + \cos^2\beta t = 1$, we can obtain the Cartesian equation of the orbits. The orbits are ellipses with centers at the origin. In this case the origin-orbit of the zero equilibrium is called a *center*.

2.10 Preliminaries of Stability

Assume that $u = 0$, the origin, is the unique equilibrium of the system

$$u' = Au,$$

which happens if and only if $\det A \neq 0$. The equilibrium $u = 0$ is said to be *stable* if small perturbations in the initial condition yield only small state deviations from the equilibrium for unlimited time. Otherwise we say that the equilibrium is *unstable*. Additionally, we say that the equilibrium $u = 0$ is *asymptotically stable* if the orbit approaches the equilibrium as $t \to +\infty$, for all nearby initial states.

The stability type of the equilibrium $u = 0$ of a planar homogeneous system with constant coefficients is easily determined by using the eigenvalues of the coefficient matrix, that is the roots of the characteristic equation

$$r^2 - (\operatorname{tr} A)r + \det A = 0.$$

According to the conclusions of the previous section the equilibrium $u = 0$ is:
(a) *asymptotically stable* when the eigenvalues are negative real numbers, or are complex conjugates with negative real part. These situations hold if and only if

$$\operatorname{tr} A < 0 \quad \text{and} \quad \det A > 0.$$

When the eigenvalues are negative real numbers we say that the origin is a *stable node*, while if the eigenvalues are complex with negative real part we say that the origin is a *stable spiral point*, or a *stable focus*.
(b) *stable*, but not asymptotically stable if

$$\operatorname{tr} A = 0 \quad \text{and} \quad \det A > 0.$$

In this situation the eigenvalues are purely imaginary, the orbits are ellipses and the origin is a *center*.
(c) *unstable* in each of these cases:

$$\operatorname{tr} A > 0 \quad \text{and} \quad \det A > 0$$

(unstable node, or unstable spiral point (unstable focus));

$$\det A < 0 \text{ (saddle point)}.$$

A more complete discussion of the stability of solutions of differential equations and systems will be given in Chapter 5. There we shall see that in general 'stability' is a property of solutions, but in the case of linear systems, it becomes a property of the system.

Chapter 3 Second-Order Differential Equations

3.1 Newton's Second Law of Motion

Just like first-order differential equations, second-order differential equations represent one of the most studied classes of differential equations. The reason is that they model numerous real processes from physics, engineering, etc. One of the most remarkable sources of second-order differential equations is Newton's second law of motion from classical mechanics, which physically involves acceleration of a particle, and mathematically the second-order derivative of position. One may say that the mechanics of the material point is essentially governed by second-order differential equations.

The general form of Newton's[1] second law is

$$mx'' = F(t, x, x'),$$

where m is the mass of the material point, $x = x(t)$ is the displacement from equilibrium at time t, and $F(t, x, x')$ is the force acting upon the material point, which may depend on time, position and velocity.

Spring-mass oscillator equation

Consider a mass m lying on a horizontal surface and connected to a spring whose second end is attached to a vertical wall. At time $t = 0$, the mass is displaced a positive distance x_0 from the equilibrium position, in the opposite sense of the vertical wall. Then the mass is released, and due to the elastic force of the spring, it oscillates back and forth around the equilibrium. Let $x(t)$ be the position of the mass at time t to the equilibrium position $x = 0$, assuming that the sense of the axis Ox is opposite to the vertical wall. Applying Newton's second law, we obtain the equation of motion

$$mx'' = F.$$

It remains to give expression to the force F. Based on experiments, it was found that in the case where the displacement was not too large, the elastic force F_e of the spring was proportional to the displacement, that is $F_e = -kx$. Here k is a positive constant depending on the elastic properties of the spring. The minus sign indicates that the force opposes the positive motion. In the case where the friction with the horizontal surface is neglected, the equation of the motion is given by Hooke's[2] law

$$mx'' = -kx.$$

---

[1] Isaac Newton (1643–1727)
[2] Robert Hooke (1635–1703)

Furthermore, if the friction force, or more generally a resistance force F_r, is considered to be proportional to the velocity, that is, $F_r = -cx'$, where c is a positive damping constant, then the equation of motion becomes

$$mx'' = -cx' - kx,$$

and is called the *damped spring-mass* equation. It is expected that in this case the amplitude of oscillations decreases in time. This equation is a homogeneous linear second-order differential equation with constant coefficients. If an additional exterior time-dependent force $f(t)$ acts on the mass, then the equation of the motion reads as

$$mx'' = -cx' - kx + f(t),$$

and is a nonhomogeneous linear second-order differential equation.

Pendulum equation

Consider a rigid pendulum of length l that is free to rotate around a fixed point, in a vertical plane, under the gravitational force. We assume that the mass of the arm of the pendulum is negligible with respect to the mass m attached at the free end of the pendulum. Let $x(t)$ be the measure of the angle that the pendulum makes with the vertical at time t. Then, the distance $s(t)$ covered by the pendulum along the circle, from the equilibrium position, is equal to the arc length, that is $lx(t)$. We assume that the motion takes place only under the action of the gravitational force $\vec{F_1}$, whose magnitude is mg, where g is the gravitational constant. This force is decomposed as

$$\vec{F_1} = \vec{F_0} + \vec{F},$$

where $\vec{F_0}$ is directed along the rigid arm of the pendulum and $\vec{F}$ is orthogonal on the pendulum and directed in the opposite sense of the motion. Since the arm of the pendulum is rigid, $\vec{F_0}$ is a passive force, hence only $\vec{F}$ is responsible for the motion. The magnitude of $\vec{F}$ being $-mg \sin x$, Newton's second law yields $m(lx)'' = -mg \sin x$, or equivalently

$$x'' + \frac{g}{l} \sin x = 0.$$

This is a second-order nonlinear differential equation called the *pendulum equation*. Here again, if the resistance to the motion takes into account and it is assumed to be proportional to the velocity, i.e. $-clx'$, then the *damped pendulum equation* is obtained, namely

$$x'' + \frac{g}{l} \sin x + \frac{c}{m} x' = 0.$$

3.2 Reduction of Order

The normal form of second-order differential equations is

$$x'' = f(t, x, x'). \tag{3.1}$$

However, some second-order equations can be reduced to first-order equations, rendering them susceptible to the simple methods of solving equations of the first order. The following are three particular types of such second-order equations:

Type 1: Second-order equations with the dependent variable missing;
Type 2: Second-order equations with the independent variable missing;
Type 3: Second-order equations with the independent variable and the derivative of the dependent variable missing.

Type 1
Consider the equations for which the dependent variable x is missing, that is

$$x'' = f(t, x').$$

It is clear that the substitution $y := x'$ yields the first-order equation

$$y' = f(t, y).$$

Example 3.1. Solve the equation $x'' = t + 2x'$.

Solution. Making the substitution $y = x'$, the equation becomes a first-order linear equation, $y' = t + 2y$. Using the method presented in Section 1.2.3 we obtain the general solution

$$y = C_1 e^{2t} - \frac{1}{2}t - \frac{1}{4}.$$

Integration finally gives the general solution of the initial equation

$$x = C_1 \frac{1}{2} e^{2t} - \frac{1}{4} t^2 - \frac{1}{4} t + C_2.$$

Type 2
Consider the equations for which the independent variable t does not appear explicitly, that is

$$x'' = f(x, x').$$

We also introduce the notation $y := x'$. Formally, from $x = x(t)$, one has $t = t(x)$, and so $y = x'(t) = x'(t(x)) = y(x)$, that is y can be considered a function of variable x. Then

$$x'' = \frac{dy}{dt} = \frac{dy}{dx} \frac{dx}{dt} = y \frac{dy}{dx},$$

and the equation becomes

$$y \frac{dy}{dx} = f(x, y),$$

a first-order equation in the unknown function $y = y(x)$. Once its solutions $y = y(x)$ have been obtained, the solutions $x = x(t)$ of the initial equation are found by solving the separable equation $x' = y(x)$.

Example 3.2. Solve the equation $x'' = 2xx'$ subject to the initial conditions $x(0) = x'(0) = 1$.

Solution. Following the method from above, we are led to the equation

$$y\frac{dy}{dx} = 2xy.$$

One solution is $y = 0$, giving $x' = 0$. Hence, in view of $x'(0) = 1$, this solution is not convenient. For $y \neq 0$, we have $dy/dx = 2x$, whose solutions are the functions $y = x^2 + C_1$. Thus, $x' = x^2 + C_1$. For $t = 0$, this gives $1 = 1 + C_1$, and so $C_1 = 0$. It remains to solve the equation $x' = x^2$. Separating the variables we obtain $x'/x^2 = 1$, where $-1/x = t + C_2$. Then $x = -(t + C_2)^{-1}$. Now the condition $x(0) = 1$ implies that $C_2 = -1$. Therefore, the solution of the Cauchy problem is $x = (1-t)^{-1}$.

Type 3

Consider the particular case of the equations of Type 2, when x' is missing, that is

$$x'' = f(x).$$

As above we arrive at a differential equation in the variables x and y,

$$y\frac{dy}{dx} = f(x),$$

which is now a separable equation. Integrating with respect to x yields

$$\frac{1}{2}y^2 = \int f(x)\, dx + E, \qquad (3.2)$$

where E is an arbitrary constant. Coming back to x', we see that (3.2) is a first-order differential equation in the separable variables t and x. Solving this equation gives the solutions of the initial equation. Note that, in the physical terms of the equation $mx'' = F(x)$, formula (3.2), written in the form

$$\frac{1}{2}my^2 - \int F(x)\, dx = E,$$

is the *energy conservation theorem*. The term $my^2/2$, where y is the velocity, is the *kinetic energy* (energy due to the motion), the term $-\int F(x)\, dx$ is the *potential energy* (energy due to external force), while the constant E represents the *total energy* in the system. Therefore, in the case where the force depends only on the position (and neither on time, nor on velocity), the total energy is conserved during the motion. Such a force is called a *conservative force*. The constant E representing the total energy can be computed if we know the initial position $x(0) = x_0$ and the initial velocity $y(0) = y_0$.

Example 3.3. A particle of mass $m = 1$ moves on the Ox axis, under the action of the conservative force $F = x$. Its initial position and velocity are $x(0) = 3$ and $x'(0) = 5$ respectively. Find the total energy of the system, write down the energy conservation theorem, and find the law of motion.

Solution. The energy conservation theorem is

$$\frac{1}{2}y^2 - \frac{1}{2}x^2 = E.$$

The amount of total energy can be found using the initial data,

$$E = \frac{1}{2}x'(0)^2 - \frac{1}{2}x(0)^2 = 8.$$

To find the law of motion we need to solve the equation given by the energy conservation theorem, namely

$$x'^2 = x^2 + 16.$$

This gives $x' = \pm\sqrt{x^2 + 16}$. Since $x'(0) = 5 > 0$, only the plus sign is convenient. Separating variables yields $\ln(x + \sqrt{x^2 + 16}) = t + C$. For $t = 0$, one has $x = 3$, and consequently $C = 3\ln 2$. Therefore, the law of motion, in the implicit form, is

$$\ln\left(x + \sqrt{x^2 + 16}\right) = t + 3\ln 2.$$

3.3 Equivalence to a First-Order System

In the previous section we saw some classes of second-order differential equations that by a change of variables can be reduced to first-order equations. Unfortunately, the reduction of order is not possible in general. However, any second-order differential equation is equivalent to a system of two first-order differential equations. Indeed, if $x \in C^2(J)$ is any solution of the equation (3.1) and we denote $y := x'$, then the couple of functions (x, y) solves the first-order differential system

$$\begin{cases} x' = y \\ y' = f(t, x, y). \end{cases} \qquad (3.3)$$

Conversely, if $(x, y) \in C^1(J, \mathbb{R}^2)$ is any solution of the system (3.3), then the first equation of the system implies that $x \in C^2(J)$, and after replacing y with x' in the second equation of the system, we find that x solves the equation (3.1). We draw attention to the very special form of the system (3.3), given by the right-hand side of its first equation.

Note that by this equivalence, the initial conditions $x(t_0) = \alpha$, $x'(t_0) = \beta$ from the statement of the Cauchy problem for the equation (3.1) correspond to the initial conditions $x(t_0) = \alpha$, $y(t_0) = \beta$ related to the equivalent system (3.3).

Additionally, by this equivalence, a linear second-order differential equation corresponds to a linear first-order differential system; a homogeneous linear equation to a homogeneous linear system; and a linear equation with constant coefficients to a linear system with constant coefficients. More exactly, the linear equation

$$x'' + a(t)x' + b(t)x = h(t)$$

is equivalent to the linear system

$$\begin{cases} x' = y \\ y' = -b(t)x - a(t)y + h(t), \end{cases}$$

and the linear equation with constant coefficients

$$x'' + ax' + bx = h(t)$$

is equivalent to the system with constant coefficients

$$\begin{cases} x' = y \\ y' = -bx - ay + h(t). \end{cases}$$

3.4 The Method of Elimination

Conversely, any coupled bidimensional linear differential system with constant coefficients and nonhomogeneous terms of class C^1, can be reduced to a second-order differential equation. Indeed, if we consider the system

$$\begin{cases} x' = a_{11}x + a_{12}y + b_1(t) \\ y' = a_{21}x + a_{22}y + b_2(t), \end{cases}$$

with $b_1 \in C^1(J)$, $b_2 \in C(J)$ and $a_{12} \neq 0$, then differentiating the first equation yields

$$x'' = a_{11}x' + a_{12}y' + b_1'(t).$$

Replacing y' by its expression given by the second equation gives

$$x'' = a_{11}x' + a_{12}(a_{21}x + a_{22}y + b_2(t)) + b_1'(t). \tag{3.4}$$

Next, from the first equation,

$$y = \frac{1}{a_{12}}\left(x' - a_{11}x - b_1(t)\right), \tag{3.5}$$

and then (3.4) becomes

$$x'' = a_{11}x' + a_{12}a_{21}x + a_{22}\left(x' - a_{11}x - b_1(t)\right) + a_{12}b_2(t) + b_1'(t).$$

Thus, the elimination of y led to the second-order linear equation with constant coefficients

$$x'' - (a_{11} + a_{22})x' + (a_{11}a_{22} - a_{12}a_{21})x = -a_{22}b_1(t) + a_{12}b_2(t) + b_1'(t).$$

Solving this equation gives x. Finally, use formula (3.5) to obtain y.

Therefore, by the elimination method, solving linear differential systems is reduced to solving higher-order linear differential equations.

3.5 Linear Second-Order Differential Equations

3.5.1 The Solution Set

The above equivalence between first-order differential systems and higher-order differential equations makes it possible for us to transfer theoretical results from linear first-order systems to linear higher-order equations. Thus we obtain the following conclusions for linear differential equations of second order with constant coefficients:

Theorem 3.4. (a) *The Cauchy problem for a linear second-order differential equation with constant coefficients has a unique solution.*
(b) *The solution set of a homogeneous linear second-order differential equation with constant coefficients is a linear space of dimension 2.*
(c) *The general solution of a linear second-order differential equation can be represented in the form*

$$x = x_h + x_p,$$

where x_h is the general solution of the homogeneous equation and x_p is a particular solution of the nonhomogeneous equation.

Note that Theorem 3.4 is also true for general linear second-order differential equations, not necessarily with constant coefficients.

3.5.2 Homogeneous Linear Equations with Constant Coefficients

As in the case of homogeneous linear systems with constant coefficients, for homogeneous linear second-order equations with constant coefficients, a basis of the solution set can be determined. Consider the equation

$$x'' + ax' + bx = 0. \tag{3.6}$$

Inspired by the case of systems, we seek solutions of the form

$$x = e^{rt}.$$

If we substitute into the equation, we obtain the condition on r, namely that it is a root of the equation

$$r^2 + ar + b = 0.$$

Notice that this equation is nothing else than the characteristic equation of the first-order differential system equivalent to the equation (3.6). We call this equation the *characteristic equation* of (3.6). Formally, it is obtained from (3.6) by replacing the derivatives of orders 0, 1 and 2, i.e. x, x' and x'', by the corresponding powers of r, that is r^0, r^1 and r^2. Three cases are possible:

1. The case of real and unequal roots

Let r_1, r_2 be the roots of the characteristic equation, and assume they are real and unequal. Then the general solution of the equation (3.6) is

$$x = C_1 e^{r_1 t} + C_2 e^{r_2 t},$$

where C_1, C_2 are arbitrary constants.

Example 3.5. Solve the equation $x'' + x' - 12x = 0$.

Solution. The characteristic equation is $r^2 + r - 12 = 0$, and its roots are $r_1 = -4, r_2 = 3$. The general solution of the differential equation is $x = C_1 e^{-4t} + C_2 e^{3t}$, where C_1, C_2 are arbitrary constants.

2. The case of equal roots

Assume that $r := r_1 = r_2$ is the double root of the characteristic equation. Then, one solution of the equation (3.6) is the function $x_1 = e^{rt}$, and a second solution is $x_2 = te^{rt}$. We can verify this by direct substitution into the equation. One has

$$x_2' = e^{rt} + rte^{rt} = (1 + rt)e^{rt},$$
$$x_2'' = re^{rt} + (1 + rt)re^{rt} = (r^2 t + 2r)e^{rt}.$$

Then

$$x_2'' + ax_2' + bx_2 = e^{rt}\left[r^2 t + 2r + a(1 + rt) + bt\right]$$
$$= e^{rt}\left[t(r^2 + ar + b) + 2r + a\right].$$

Since r is the double root of the characteristic equation, one has $r^2 + ar + b = 0$ and $2r + a = 0$. Hence $x_2'' + ax_2' + bx_2 = 0$, as claimed. Thus, in this case, the general solution of (3.6) is

$$x = C_1 e^{rt} + C_2 t e^{rt},$$

where C_1, C_2 are arbitrary constants.

Example 3.6. Solve the equation $x'' + 4x' + 4x = 0$.

Solution. The characteristic equation is $r^2 + 4r + 4 = 0$, with double root $r = -2$. Hence the general solution of the differential equation is $x = C_1 e^{-2t} + C_2 t e^{-2t}$, where C_1, C_2 are arbitrary constants.

3. The case of complex conjugate roots

Assume that $\alpha \pm \beta i$ are the complex conjugate roots of the characteristic equation. Then the complex-valued functions $e^{(\alpha + i\beta)t}$ and $e^{(\alpha - i\beta)t}$, or more precisely

$$e^{\alpha t}(\cos\beta t + i\sin\beta t), \quad e^{\alpha t}(\cos\beta t - \sin\beta t),$$

are complex-valued solutions of the differential equation (3.6). Since the equation is linear and homogeneous, a linear combination with complex coefficients of any two solutions is also a solution. In particular, the functions

$$\text{Re } e^{(\alpha+i\beta)t} = \frac{1}{2}\left(e^{(\alpha+i\beta)t} + e^{(\alpha-i\beta)t}\right) = e^{\alpha t}\cos\beta t,$$

$$\text{Im } e^{(\alpha+i\beta)t} = \frac{1}{2i}\left(e^{(\alpha+i\beta)t} - e^{(\alpha-i\beta)t}\right) = e^{\alpha t}\sin\beta t$$

are solutions of the differential equation. Thus the general solution of (3.6) is

$$x = C_1 e^{\alpha t}\cos\beta t + C_2 e^{\alpha t}\sin\beta t,$$

where C_1, C_2 are arbitrary constants.

Example 3.7. Solve the equation $x'' + 6x' + 34x = 0$.

Solution. The characteristic equation is $r^2 + 6r + 34 = 0$, with the roots $-3 \pm 5i$. Thus the general solution of the differential equation is $x = e^{-3t}(C_1 \cos 5t + C_2 \sin 5t)$, where C_1, C_2 are arbitrary constants.

3.5.3 Variation of Parameters Method

Using two linearly independent solutions x_1, x_2 of the homogeneous equation (3.6) we are able to find a particular solution x_p of the nonhomogeneous equation

$$x'' + ax' + bx = h(t) \tag{3.7}$$

by means of the method of *variation of parameters*. In this case, we try a solution x_p of the form

$$x_p = C_1(t)x_1 + C_2(t)x_2.$$

Then

$$x_p' = C_1'(t)x_1 + C_2'(t)x_2 + C_1(t)x_1' + C_2(t)x_2'.$$

In order to simplify the expression of x_p', we require that

$$C_1'(t)x_1 + C_2'(t)x_2 = 0,$$

so

$$x_p' = C_1(t)x_1' + C_2(t)x_2'.$$

Furthermore

$$x_p'' = C_1'(t)x_1' + C_2'(t)x_2' + C_1(t)x_1'' + C_2(t)x_2'',$$

and substituting in (3.7) yields

$$C_1(t)(x_1'' + ax_1' + bx_1) + C_2(t)(x_2'' + ax_2' + bx_2)$$
$$+ C_1'(t)x_1' + C_2'(t)x_2' = h(t).$$

Since x_1 and x_2 are solutions of the homogeneous equation, the two expressions in brackets are zero, so what remains is

$$C'_1(t)x'_1 + C'_2(t)x'_2 = h(t).$$

Thus the calculation of the variable parameters $C_1(t)$ and $C_2(t)$ is reduced to solving the algebraic system

$$\begin{cases} C'_1(t)x_1 + C'_2(t)x_2 = 0 \\ C'_1(t)x'_1 + C'_2(t)x'_2 = h(t). \end{cases} \quad (3.8)$$

This system is compatible since its determinant

$$\begin{vmatrix} x_1 & x_2 \\ x'_1 & x'_2 \end{vmatrix}$$

is nonzero as a consequence of the linear independence of the two solutions x_1, x_2.

Example 3.8. Solve the equation

$$x'' - 3x' + 2x = -e^t,$$

if one knows that $x_1 = e^t$ and $x_2 = e^{2t}$ are two particular solutions of the corresponding homogeneous equation.

Solution. Use the method of variation of parameters to obtain a particular solution of the nonhomogeneous equation, in the form $x_p = C_1(t)e^t + C_2(t)e^{2t}$. Solving (3.8), that is

$$\begin{cases} C'_1(t)e^t + C'_2(t)e^{2t} = 0 \\ C'_1(t)e^t + 2C'_2(t)e^{2t} = -e^t, \end{cases}$$

yields $C'_1(t) = 1$, $C'_2(t) = -e^{-t}$. Hence $C_1(t) = t$ and $C_2(t) = e^{-t}$. Thus we have found the particular solution $x_p = (t+1)e^t$. The general solution of the equation is

$$x = C_1 e^t + C_2 e^{2t} + t e^t,$$

where C_1, C_2 are arbitrary constants.

3.5.4 The Method of Undetermined Coefficients

For some special right-hand sides $h(t)$ of the equation (3.7), instead of the general method of variation of parameters one can use the *method of undetermined coefficients*. It consists of looking for a solution under the form of the right-hand side. Below are some basic cases when this method can be used.

(1) If $h(t) = P_n(t) := a_n t^n + a_{n-1} t^{n-1} + \ldots + a_1 t + a_0$ is a polynomial of degree n, then we try a solution x_p of the same type, i.e.

$$x_p = Q_n(t) := A_n t^n + A_{n-1} t^{n-1} + \ldots + A_1 t + A_0,$$

where the coefficients $A_0, A_1, \ldots, A_n$ are found by substituting x_p into (3.7).

3.5 Linear Second-Order Differential Equations — 63

Example 3.9. Find a particular solution of the equation

$$x'' + 7x' + 12x = 3t^2 - 4.$$

Solution. The right-hand side, $h(t) = 3t^2 - 4$, is a polynomial of degree 2, so we try $x_p = at^2 + bt + c$. We have $x'_p = 2at + b$ and $x''_p = 2a$. Substituting gives

$$2a + 14at + 7b + 12at^2 + 12bt + 12c = 3t^2 - 4.$$

Equating coefficients gives $12a = 3$, $14a + 12b = 0$ and $2a + 7b + 12c = -4$. Therefore $a = 1/4$, $b = -7/24$ and $c = -59/288$. Thus a particular solution of the equation is

$$x_p = \frac{1}{4}t^2 - \frac{7}{24}t - \frac{59}{288}.$$

(2) If $h(t) = P_n(t)e^{\gamma t}$, then we try a solution x_p of the same type, i.e.

$$x_p = Q_n(t)e^{\gamma t}.$$

Example 3.10. Find a particular solution of the equation

$$x'' + 7x' + 12x = 21e^{3t}.$$

Solution. We try $x_p = ae^{3t}$. Substituting into the equation gives $a = 1/2$. Thus $x_p = e^{3t}/2$.

(3) If $h(t) = P_n(t)e^{\gamma t}\sin \omega t$, or $h(t) = P_n(t)e^{\gamma t}\cos \omega t$, then a particular solution is sought in the form

$$x_p = e^{\gamma t}(Q_n(t)\sin \omega t + R_n(t)\cos \omega t).$$

Substituting into the equation and equating like terms gives the coefficients of the polynomials $Q_n(t)$ and $R_n(t)$. Notice that, even if h contains only one of the functions sin and cos, the determination of x_p only in sin, or only in cos, is in general impossible. The reason is that by differentiation, the function sin turns into cos, and vice versa.

Example 3.11. Find a particular solution of the equation

$$x'' - x' + x = 61e^{3t}\sin t.$$

Solution. We try a solution of the form $x_p = e^{3t}(a\sin t + b\cos t)$. One has $x'_p = e^{3t}[(3a - b)\sin t + (a + 3b)\cos t]$ and $x''_p = e^{3t}[(8a - 6b)\sin t + (6a + 8b)\cos t]$. Substituting and equating like terms gives $6a - 5b = 61$ and $5a + 6b = 0$. Hence $a = 6$, $b = -5$, and a particular solution is $x_p = e^{3t}(6\sin t - 5\cos t)$.

There are situations covered by the above cases when the identification of the coefficients is nonetheless impossible. Then the guess of the particular solution requires some corrections. Below are these situations and the necessary corrections that need to be done.

(1') If $h(t) = P_n(t)$ and $r = 0$ is a root of the characteristic equation of multiplicity m, then we try a particular solution in the form

$$x_p = t^m Q_n(t).$$

Example 3.12. Find a particular solution of the equation

$$x'' - 6x' = 3t - 5.$$

Solution. The characteristic equation is $r^2 - 6r = 0$ and $r = 0$ is a root of multiplicity 1. Thus we try $x_p = t(at + b)$. Substituting gives $x_p = -t^2/4 + 3t/4$.

(2') If $h(t) = P_n(t)e^{\gamma t}$ and $r = \gamma$ is a root of the characteristic equation of multiplicity m, then we try

$$x_p = t^m Q_n(t) e^{\gamma t}.$$

Example 3.13. Give the form of a particular solution of the equation

$$x'' - 2x' + x = (t - 8)e^t.$$

Solution. The form of the particular solution is $x_p = t^2(at + b)e^t$, since $r = 1$ is a root of the characteristic equation of multiplicity 2.

(3') If $h(t) = P_n(t)e^{\gamma t} \sin \omega t$, or $h(t) = P_n(t)e^{\gamma t} \cos \omega t$, and $r = \gamma + \omega i$ is a root of the characteristic equation of multiplicity m, then we try a particular solution in the form

$$x_p = t^m e^{\gamma t} (Q_n(t) \sin \omega t + R_n(t) \cos \omega t).$$

Example 3.14. Give the form of a particular solution of the equation

$$x'' - 6x' + 34x = e^{3t}(t + 2) \cos 5t.$$

Solution. The number $3 + 5i$ is a simple root of the characteristic equation. Thus, $x_p = te^{3t}[(at + b) \sin 5t + (ct + d) \cos 5t]$.

3.5.5 Euler Equations

Compared to linear differential equations with constant coefficients, which are explicitly solvable, linear equations with variable coefficients rarely have explicit solutions. However, for a special class of linear equations with variable coefficients, namely the Euler equations, explicit solutions are easy to obtain by changes of variable leading to linear equations with constant coefficients. The second-order *Euler*[3] equation is

$$t^2 x'' + atx' + bx = h(t).$$

As main feature of Euler equations, we highlight that the variable coefficients are powers of t having the exponent equal to the order of the corresponding derivative of x.

---

[3] Leonhard Euler (1707–1783)

Assume that the above equation is considered for $t > 0$, and make the change of variable

$$s = \ln t.$$

One has

$$x' = \frac{dx}{dt} = \frac{dx}{ds}\frac{ds}{dt} = \frac{1}{t}\frac{dx}{ds},$$

$$x'' = \frac{d}{dt}\left(\frac{1}{t}\frac{dx}{ds}\right) = -\frac{1}{t^2}\frac{dx}{ds} + \frac{1}{t}\frac{d}{dt}\left(\frac{dx}{ds}\right)$$

$$= -\frac{1}{t^2}\frac{dx}{ds} + \frac{1}{t}\frac{d^2x}{ds^2}\frac{ds}{dt} = -\frac{1}{t^2}\frac{dx}{ds} + \frac{1}{t^2}\frac{d^2x}{ds^2}.$$

Substituting into the equation yields a linear equation with constant coefficients,

$$\frac{d^2x}{ds^2} + (a-1)\frac{dx}{ds} + bx = H(s),$$

where $H(s) = h(e^s)$.

Example 3.15. Solve the equation

$$t^2 x'' + 2tx' - 12x = 6\ln t, \quad t > 0.$$

Solution. Making the change of variable $s = \ln t$, the equation becomes

$$x'' + x' - 12x = 6s,$$

where now $x = x(s)$. The general solution of this equation with constant coefficients is

$$x(s) = C_1 e^{-4s} + C_2 e^{3s} - \frac{1}{2}s - \frac{1}{24}.$$

Coming back to the variable t, we obtain the general solution of the Euler equation

$$x = C_1 \frac{1}{t^4} + C_2 t^3 - \frac{1}{2}\ln t - \frac{1}{24},$$

where C_1, C_2 are arbitrary constants.

Note that in the case of homogeneous Euler equations, we may directly find solutions of the form

$$x = t^\alpha,$$

where the exponent α is determined by substituting t^α into the equation.

Example 3.16. Solve the homogeneous Euler equation

$$t^2 x'' + 2tx' - 12x = 0.$$

Solution. Letting $x = t^\alpha$ yields the equation $\alpha(\alpha - 1) + 2\alpha - 12 = 0$, which has the roots $\alpha_1 = -4$ and $\alpha_2 = 3$. These provide two linearly independent solutions $x_1 = t^{-4}$ and $x_2 = t^3$. Thus the general solution of the Euler equation is

$$x = C_1 \frac{1}{t^4} + C_2 t^3,$$

where C_1, C_2 are arbitrary constants.

3.6 Boundary Value Problems

We already know that the general solution of any linear second-order differential equation depends on two arbitrary constants C_1 and C_2. We also know that the solution of the Cauchy problem relative to such an equation is unique, which means that the two constants are completely determined by the initial conditions. To select a particular solution from the general solution, that is, to determine the constants C_1 and C_2, is also possible by means of some other conditions than the initial ones. Usually, these conditions are stated in terms of the values of the functions x and x' at the ends of a given interval; let it be $[0, 1]$, for simplicity. Such conditions are called *bilocal*, or *boundary conditions*, and are inspired by real constraints from concrete physical problems. In what follows we shall restrict ourselves to boundary conditions of the type

$$\alpha_1 x(0) + \alpha_2 x'(0) = 0, \tag{3.9}$$
$$\beta_1 x(1) + \beta_2 x'(1) = 0,$$

where $\alpha_1, \alpha_2, \beta_1, \beta_2$ are real coefficients, and we shall refer to the linear equation (3.7) for $t \in [0, 1]$. Thus, we consider the *boundary value problem*

$$\begin{cases} x'' + ax' + bx = h(t), & t \in [0, 1] \\ \alpha_1 x(0) + \alpha_2 x'(0) = 0 \\ \beta_1 x(1) + \beta_2 x'(1) = 0, \end{cases} \tag{3.10}$$

where $a, b, \alpha_1, \alpha_2, \beta_1, \beta_2 \in \mathbb{R}$ and $h \in C[0, 1]$.

Let us assume that the zero function is the unique solution $x \in C^2[0, 1]$ of the homogeneous problem

$$\begin{cases} x'' + ax' + bx = 0, & t \in [0, 1] \\ \alpha_1 x(0) + \alpha_2 x'(0) = 0 \\ \beta_1 x(1) + \beta_2 x'(1) = 0. \end{cases} \tag{3.11}$$

Then, for each $h \in C[0, 1]$, problem (3.10) has at most one solution (we leave the proof of this statement to the reader). One can prove the existence of the solution and give its integral representation in terms of the function h. To this aim, consider two nonzero solutions ϕ and ψ of the homogeneous equation $x'' + ax' + bx = 0$ such that ϕ satisfies the first boundary condition, while ψ satisfies the second boundary condition, i.e.

$$\alpha_1 \phi(0) + \alpha_2 \phi'(0) = 0,$$
$$\beta_1 \psi(1) + \beta_2 \psi'(1) = 0.$$

Also consider the determinant $W(t)$, called the *Wronskian*[4] of these two solutions,

$$W(t) = \begin{vmatrix} \phi(t) & \psi(t) \\ \phi'(t) & \psi'(t) \end{vmatrix}.$$

It is easy to see that

$$W(t) \neq 0 \quad \text{for every } t \in [0, 1].$$

Indeed, otherwise, there is some $t_0 \in [0, 1]$ with $W(t_0) = 0$. The columns of the determinant are then proportional, that is, there exist constants y and η such that $y^2 + \eta^2 \neq 0$,

$$y\phi(t_0) + \eta\psi(t_0) = 0 \quad \text{and} \quad y\phi'(t_0) + \eta\psi'(t_0) = 0.$$

This shows that the function $y\phi + \eta\psi$ is the solution of the Cauchy problem

$$\begin{cases} x'' + ax' + bx = 0, & t \in [0, 1] \\ x(t_0) = x'(t_0) = 0. \end{cases}$$

Consequently, $y\phi + \eta\psi$ is the zero function. If $y \neq 0$, then $\phi = -(\eta/y)\psi$ satisfies (3.11), so $\phi = 0$, which is impossible. Similarly we derive a contradiction if $\eta \neq 0$. Thus $W(t) \neq 0$ for every $t \in [0, 1]$.

Define the function $G: [0, 1] \times [0, 1] \to \mathbb{R}$,

$$G(t, s) = \begin{cases} -\frac{\phi(t)\psi(s)}{W(s)}, & 0 \leq t \leq s \leq 1 \\ -\frac{\phi(s)\psi(t)}{W(s)}, & 0 \leq s < t \leq 1. \end{cases} \tag{3.12}$$

Then the function

$$x(t) = -\int_0^1 G(t, s)h(s)\, ds, \quad t \in [0, 1]$$

is the solution of the boundary value problem (3.10). Indeed, based on the following formulas:

$$x(t) = \psi(t) \int_0^t \frac{\phi(s)}{W(s)} h(s)\, ds + \phi(t) \int_t^1 \frac{\psi(s)}{W(s)} h(s)\, ds,$$

$$x'(t) = \psi'(t) \int_0^t \frac{\phi(s)}{W(s)} h(s)\, ds + \phi'(t) \int_t^1 \frac{\psi(s)}{W(s)} h(s)\, ds,$$

---

[4] Józef Maria Hoene-Wroński (1776–1853)

$$x''(t) = \psi''(t) \int_0^t \frac{\phi(s)}{W(s)} h(s)\, ds + \phi''(t) \int_t^1 \frac{\psi(s)}{W(s)} h(s)\, ds$$
$$+ \frac{\psi'(t)\phi(t)}{W(t)} h(t) - \frac{\phi'(t)\psi(t)}{W(t)} h(t)$$
$$= \psi''(t) \int_0^t \frac{\phi(s)}{W(s)} h(s)\, ds + \phi''(t) \int_t^1 \frac{\psi(s)}{W(s)} h(s)\, ds + h(t),$$

we can easily verify that x is the solution of (3.10).

The function G is called the *Green's[5] function* of the boundary value problem.

It is remarkable that the construction of Green's function does not depend on the choice of the two solutions ϕ and ψ. Indeed, if $G_1(t, s)$ is the function defined on the model of (3.12), using other two solutions ϕ_1, ψ_1, then for the solution x of the boundary value problem (3.10) we have the representations

$$x(t) = -\int_0^1 G(t, s) h(s)\, ds = -\int_0^1 G_1(t, s) h(s)\, ds \quad (t \in [0, 1]),$$

where

$$\int_0^1 (G(t, s) - G_1(t, s)) h(s)\, ds = 0$$

for all $t \in [0, 1]$ and $h \in C[0, 1]$. For a fixed but arbitrary t, choose $h(s) = G(t, s) - G_1(t, s)$ to obtain

$$\int_0^1 (G(t, s) - G_1(t, s))^2\, ds = 0.$$

This yields $G(t, s) - G_1(t, s) = 0$ for all $s \in [0, 1]$. Consequently, $G_1(t, s) = G(t, s)$ for all $t, s \in [0, 1]$, as claimed.

Example 3.17. Construct Green's function of the problem $x'' = h(t)$, $t \in [0, 1]$; $x(0) = x(1) = 0$.

Solution. The solutions of the homogeneous equation $x'' = 0$ are $x = C_1 + C_2 t$. Choose ϕ, one of the solutions satisfying $x(0) = 0$, for example $\phi = t$, and choose ψ, one of the solutions satisfying $x(1) = 0$, for example $\psi = 1 - t$. Then

$$W(t) = \begin{vmatrix} t & 1-t \\ 1 & -1 \end{vmatrix} = -t - 1 + t = -1,$$

and thus

$$G(t, s) = \begin{cases} t(1 - s), & 0 \le t \le s \le 1 \\ s(1 - t), & 0 \le s < t \le 1. \end{cases}$$

---

5 George Green (1793–1841)

Example 3.18. Find Green's function of the problem $x'' - x = h(t)$, $t \in [0, 1]$; $x'(0) = x'(1) = 0$.

Solution. Choose $\phi = \cosh t$ and $\psi = \cosh(1 - t)$. Then $W(t) = -\sinh 1$ and thus

$$G(t, s) = \begin{cases} \dfrac{\cosh t \cosh(1 - s)}{\sinh 1}, & 0 \le t \le s \le 1 \\ \dfrac{\cosh s \cosh(1 - t)}{\sinh 1}, & 0 \le s < t \le 1. \end{cases}$$

3.7 Higher-Order Linear Differential Equations

The theory of second-order differential equations can be easily extended to differential equations of any order n. Thus, the Cauchy problem related to the equation

$$x^{(n)} = f\left(t, x, x', \ldots, x^{(n-1)}\right) \tag{3.13}$$

is stated by means of n conditions

$$x(t_0) = \alpha_0, x'(t_0) = \alpha_1, \ldots, x^{(n-1)}(t_0) = \alpha_{n-1}.$$

Also note that the n-order equation (3.13) is equivalent, via the notations

$$x_1 = x, \quad x_2 = x', \quad \ldots, \quad x_n = x^{(n-1)},$$

to the n-dimensional first-order differential system

$$\begin{cases} x_1' = x_2 \\ x_2' = x_3 \\ \ldots \\ x_{n-1}' = x_n \\ x_n' = f(t, x_1, x_2, \ldots, x_n). \end{cases} \tag{3.14}$$

The system (3.14) is linear if the equation (3.13) is linear, that is of the form

$$x^{(n)} + a_1(t)x^{(n-1)} + \ldots + a_n(t)x = h(t).$$

Additionally, the system (3.14) is linear with constant coefficients if the equation (3.13) is, i.e. it has the form

$$x^{(n)} + a_1 x^{(n-1)} + \ldots + a_n x = h(t). \tag{3.15}$$

Consequently, Theorem 3.4 holds with 2 replaced by n for linear differential equations of any order n.

As in the case of second-order equations, in order to solve the n-order homogeneous linear equation with constant coefficients,

$$x^{(n)} + a_1 x^{(n-1)} + \ldots + a_n x = 0,$$

we seek solutions of the form $x = e^{rt}$. After substituting into the equation, we find for r the condition of being a root of the characteristic equation

$$r^n + a_1 r^{n-1} + \ldots + a_{n-1} r + a_n = 0.$$

To any real root r of the characteristic equation of multiplicity m, there correspond m linearly independent solutions of the homogeneous differential equation

$$e^{rt}, \quad te^{rt}, \quad t^2 e^{rt}, \quad \ldots, \quad t^{m-1} e^{rt}.$$

To any pair of complex conjugate roots $\alpha \pm \beta i$ of multiplicity m, there correspond $2m$ linearly independent solutions of the homogeneous differential equation

$$e^{\alpha t} \cos \beta t, \, te^{\alpha t} \cos \beta t, \, t^2 e^{\alpha t} \cos \beta t, \ldots, t^{m-1} e^{\alpha t} \cos \beta t,$$
$$e^{\alpha t} \sin \beta t, \, te^{\alpha t} \sin \beta t, \, t^2 e^{\alpha t} \sin \beta t, \ldots, t^{m-1} e^{\alpha t} \sin \beta t.$$

Furthermore, if $x_1, x_2, \ldots, x_n$ are n linearly independent solutions of the homogeneous equation associated with (3.15), then a particular solution of the nonhomogeneous equation can be determined using the method of variation of parameters, in the form

$$x_p = C_1(t) x_1 + C_2(t) x_2 + \ldots + C_n(t) x_n,$$

where the variable coefficients $C_i(t)$, $i = 1, 2, \ldots, n$ are obtained, solving by Cramer's rule the algebraic linear system

$$\begin{cases} C_1'(t) x_1 + C_2'(t) x_2 + \ldots + C_n'(t) x_n = 0 \\ C_1'(t) x_1' + C_2'(t) x_2' + \ldots + C_n'(t) x_n' = 0 \\ \ldots \\ C_1'(t) x_1^{(n-2)} + C_2'(t) x_2^{(n-2)} + \ldots + C_n'(t) x_n^{(n-2)} = 0 \\ C_1'(t) x_1^{(n-1)} + C_2'(t) x_2^{(n-1)} + \ldots + C_n'(t) x_n^{(n-1)} = h(t). \end{cases}$$

Additionally, the method of undetermined coefficients is applicable to general n-order nonhomogeneous equations with constant coefficients, as in the situations presented in Section 3.5.4.

Finally, the Euler equations of order n are of the form

$$t^n x^{(n)} + a_1 t^{n-1} x^{(n-1)} + \ldots + a_{n-1} t x' + a_n x = h(t)$$

and can be reduced to equations with constant coefficients making the change of variable $s = \ln t$.

Example 3.19. Solve the equation

$$x^{(IV)} + x''' - 3x'' - 5x' - 2x = -90e^{-t} - 90. \qquad (3.16)$$

Solution. The characteristic equation is $r^4 + r^3 - 3r^2 - 5r - 2 = 0$ and its roots are $r = -1$ of multiplicity $m = 3$, and $r = 2$ of multiplicity 1. It follows that the general solution of the associated homogeneous equation is

$$x_h = (C_1 + C_2 t + C_3 t^2)e^{-t} + C_4 e^{2t}.$$

Let Lx denote the differential expression from the left-hand side of the equation, i.e.

$$Lx = x^{(IV)} + x''' - 3x'' - 5x' - 2x.$$

A particular solution of the equation

$$Lx = -90$$

is sought of the form $x_{p_1} = a$ (a constant like the right-hand side of the equation). We obtain $a = 45$. Next we try a particular solution of the equation

$$Lx = -90e^{-t}$$

in the form $x_{p_2} = bt^3 e^{-t}$. Here be^{-t} reproduces the form of the right-hand side of the equation, while the power t^3 is needed since -1 is a triple root of the characteristic equation. Substituting into the equation gives $b = 5$. Thus a particular solution of (3.16) is $x_p = x_{p_1} + x_{p_2} = 45 + 5t^3 e^{-t}$, and the general solution is

$$x = (C_1 + C_2 t + C_3 t^2)e^{-t} + C_4 e^{2t} + 45 + 5t^3 e^{-t},$$

where C_1, C_2, C_3, C_4 are arbitrary constants.

Chapter 4 Nonlinear Differential Equations

As we have seen so far, there is a unitary theory of linear differential equations and systems essentially based on linear algebra. In contrast, the study of nonlinear differential equations and systems that arise most frequently in mathematical modeling is much more complex. The goal of this chapter is to present some elements of qualitative analysis of nonlinear differential equations and systems. Since differential equations of any order are equivalent to first-order differential systems, it suffices to focus the analysis on the latter. Therefore, we shall consider first-order equations of the form

$$x' = f(t, x),$$

two-dimensional (or planar) systems

$$\begin{cases} x' = f(t, x, y) \\ y' = g(t, x, y), \end{cases}$$

and, generally, n-dimensional systems of the type (2.1), represented under the form of a vector equation,

$$u' = F(t, u).$$

4.1 Mathematical Models Expressed by Nonlinear Systems

In Chapter 1 we presented some nonlinear differential equations modeling real processes from physics and biology. In all these models, there was only one fundamental variable representing the state of the process at any time. More complex phenomena need more than one fundamental variable in order to be described. In this case, the mathematical models are expressed by systems of differential equations.

Here are some examples of nonlinear differential systems modeling biological processes.

4.1.1 The Lotka–Volterra Model

We present a model describing the interaction between a predator species and a prey one. Let $x(t)$ be the prey density and $y(t)$ be the predator density at time t. In the absence of predators, there is no inhibition in the growth of the prey, which follows Malthus' law $x' = rx$, increasing exponentially in time. Here r is the per capita growth rate. Analogously, in the absence of prey, the predators are expected to decrease exponentially in time according to the law $y' = -my$, where m is per capita death rate. The interaction between the two species is kept in equilibrium by those terms that contribute to the decrease of the prey population and the increase of the predator population. We may accept that the per capita growth rate of the prey population diminishes proportionally with predator density, that is, $x'/x = r - ay$, and similarly, that the predator per capita rate is ameliorated proportionally with the prey density, hence $y'/y = -m+bx$. Thus we obtain the simplest model of predator-prey interaction, called the Lotka[1]–Volterra[2] system:

$$\begin{cases} x' = x(r - ay) \\ y' = -y(m - bx) . \end{cases}$$

From the first equation, we observe that as long as the predator density is smaller than r/a, one has $x' > 0$, so the prey population increases. On the contrary, during those periods of time when y exceeds the threshold r/a, one has $x' < 0$, and the prey population decreases. Similar remarks can be made using the second equation, on the fluctuation of predator density.

The model can be refined in order to incorporate some other aspects such as migration and overexploitation.

4.1.2 The SIR Epidemic Model

Assume that a closed fixed population of size N is affected by a disease that spreads by contact with infective individuals. During the epidemic, at any time, the population divides into three classes: the *susceptible* class of those individuals who can get the illness, but not yet infected; the *infective* class, of those that are infected; and the *removed* class, of those who cannot get the illness because they have recovered or are immune to the virus. Let $S(t)$, $I(t)$ and $R(t)$ denote the number of individuals in each of the classes, at time t. It is clear that $R(t) = N - S(t) - I(t)$. If a is the *transmission coefficient* of the disease, then the per capita rate of the diminishing susceptible population is proportional with the number of infected individuals, i.e. $-aI(t)$. Hence $S'/S = -aI$, or $S' = -aSI$. The quantity aSI is the infection rate, or the number of individuals who at time t go from the susceptible class S into the infective class I. Then $I' = aSI - rI$, where rI is the removal rate, or the population going from the infective class I, into the removed class R. The proportionality factor r is called the *recovery rate*. Thus we obtain the simplest model for epidemics, called the *SIR model*:

$$\begin{cases} S' = -aSI \\ I' = aSI - rI . \end{cases}$$

---

[1] Alfred James Lotka (1880–1949)
[2] Vito Volterra (1860–1940)

If the epidemiological state of the population is known at the initial time $t = 0$, that is the initial values are known,
$$S(0) = S_0, \quad I(0) = I_0,$$
then by solving the corresponding Cauchy problem we can predict the evolution of the disease.

The model can be modified in order to incorporate some other aspects related to epidemics, such as vaccination, population age structure, or the possibility of reinfection.

4.1.3 An Immunological Model

We present a model for the interaction inside the body between a virus and the immune system. Let $V(t)$ be the number of virions in an organism, $X(t)$ the number of uninfected target cells, and $Y(t)$ the number of infected cells. First, neglecting the immune response, we may state the model
$$\begin{cases} V' = aY - bV \\ X' = c - dX - eXV \\ Y' = eXV - gY. \end{cases}$$
These equalities are understood as follows:
- The growth rate of the virus population is proportional to the number Y of infected cells, and the virus specific death rate is b;
- The uninfected cells are uniformly produced by the organism at a rate c, die at a specific rate d, and are infected by virus at a specific rate eV, proportional to the number V of virions;
- The cells in the X class give the growth rate for the infected cells that die at a specific rate $g = d + f$, where d is the natural death rate and f is the additional death rate caused by the infection.

Next, we take into consideration the response of the immune system, denoting by $Z(t)$ the number of killer cells produced by the organism for the elimination of infected cells. Assume that the killer cells are produced at a constant rate h and die at a specific rate k. Also assume that killer cells eliminate infected cells at a rate mYZ proportional to both Y and Z. Thus the model becomes
$$\begin{cases} V' = aY - bV \\ X' = c - dX - eXV \\ Y' = eXV - gY - mYZ \\ Z' = h - kZ. \end{cases}$$
This model can be modified in order to describe the AIDS disease caused by the HIV virus, which infects the cells of the immune system themselves.

4.1.4 A Model in Hematology

Most blood cells in the body are produced in bone marrow, from primitive stem cells that have the capacity of self-renewal and differentiation. Genetic alterations of stem cells can lead to the abnormal proliferation of mutant cells, called leukemic cells, which become dominant in the competition with normal cells. The dynamics of the normal and leukemic cell populations can be described by the following differential system:

$$\begin{cases} x' = \frac{ax}{1+b(x+y)} - cx \\ y' = \frac{Ay}{1+B(x+y)} - Cy, \end{cases} \quad (4.1)$$

where $x(t)$ and $y(t)$ represent the number of normal and leukemic cells respectively at time t; a, A are the growth rates of the two cell populations; c, C are the cell death rates; and b, B stand for the sensibility of the two types of cells with respect to the crowding effect in the bone marrow microenvironment. Let us note the inhibitory role of the factors $[1 + b(x + y)]^{-1}$ and $[1 + B(x + y)]^{-1}$ in the situation that the total cell population $x + y$ becomes large.

4.2 Gronwall's Inequality

We start our qualitative analysis of nonlinear differential equations with a useful tool for the study of the Cauchy problem, namely Gronwall's[3] inequality.

Let J be an interval of any type of the real line.

Theorem 4.1 ('Right' Gronwall's inequality). *Let $\phi \in C(J, \mathbb{R})$, $a \in \mathbb{R}$, $b \in \mathbb{R}_+$ and $t_0 \in J$. If*

$$\phi(t) \le a + b \int_{t_0}^{t} \phi(s)\, ds$$

for all $t \in [t_0, +\infty) \cap J$, then

$$\phi(t) \le a e^{b(t-t_0)}$$

for every $t \in [t_0, +\infty) \cap J$.

Proof. Denote

$$\psi(t) := a + b \int_{t_0}^{t} \phi(s)\, ds\,.$$

From the hypothesis, one has

$$\phi(t) \le \psi(t) \quad (4.2)$$

---

[3] Thomas Hakon Gronwall (1877–1932)

for all $t \in [t_0, +\infty) \cap J$. Differentiating gives
$$\psi'(t) = b\phi(t).$$
Since b is nonnegative, from (4.2) one has $b\phi(t) \le b\psi(t)$. Hence
$$\psi'(t) \le b\psi(t).$$
Multiplying by $e^{-b(t-t_0)}$ yields
$$\psi'(t)e^{-b(t-t_0)} - b\psi(t)e^{-b(t-t_0)} \le 0,$$
or equivalently
$$\left(\psi(t)e^{-b(t-t_0)}\right)' \le 0.$$
Integrating on the interval $[t_0, t]$, where $t \in J$, $t_0 < t$, gives
$$\psi(t)e^{-b(t-t_0)} - \psi(t_0) \le 0.$$
Then, since $\psi(t_0) = a$,
$$\psi(t) \le \psi(t_0)e^{b(t-t_0)} = ae^{b(t-t_0)}.$$
Finally use again (4.2) to obtain the conclusion
$$\phi(t) \le ae^{b(t-t_0)}. \qquad \square$$

A 'left' version of Gronwall's inequality also holds. Its proof remains as a good exercise for the reader.

Theorem 4.2 ('Left' Gronwall's inequality). *Let $\phi \in C(J, \mathbb{R})$, $a \in \mathbb{R}$, $b \in \mathbb{R}_+$ and $t_0 \in J$. If*
$$\phi(t) \le a + b \int_t^{t_0} \phi(s)\,ds$$
for all $t \in (-\infty, t_0] \cap J$, then
$$\phi(t) \le ae^{b(t_0-t)}$$
for every $t \in (-\infty, t_0] \cap J$.

The 'right' and 'left' versions of Gronwall's inequality yield the 'bilateral' version of Gronwall's inequality.

Theorem 4.3 (Gronwall's inequality). *Let $\phi \in C(J, \mathbb{R}_+)$; $a, b \in \mathbb{R}_+$ and $t_0 \in J$. If*
$$\phi(t) \le a + b \left| \int_{t_0}^t \phi(s)\,ds \right|$$
for all $t \in J$, then
$$\phi(t) \le ae^{b|t-t_0|}$$
for every $t \in J$.

Proof. Since ϕ is nonnegative on J, for $t \in [t_0, +\infty) \cap J$ one has

$$\left| \int_{t_0}^t \phi(s)\, ds \right| = \int_{t_0}^t \phi(s)\, ds ,$$

and the conclusion follows from the 'right' version of Gronwall's inequality, while for $t \in (-\infty, t_0] \cap J$ one has

$$\left| \int_{t_0}^t \phi(s)\, ds \right| = \int_t^{t_0} \phi(s)\, ds ,$$

and the conclusion follows from the 'left' version of Gronwall's inequality. □

The following immediate consequence of Gronwall's inequality is particularly useful for applications:

Corollary 4.4. *Let $\phi \in C(J, \mathbb{R}_+)$, $b \in \mathbb{R}_+$ and $t_0 \in J$. If*

$$\phi(t) \le b \left| \int_{t_0}^t \phi(s)\, ds \right|$$

for all $t \in J$, then $\phi(t) = 0$ for every $t \in J$.

In the following sections we present two applications of Gronwall's inequality to the uniqueness and continuous dependence on the initial value of the solution of the Cauchy problem. First we discuss the Cauchy problem for a first-order differential equation, and then, briefly, the Cauchy problem for first-order differential systems and for higher-order differential equations.

4.3 Uniqueness of Solutions for the Cauchy Problem

Let $f: D \to \mathbb{R}$ be a function defined on a set $D \subset \mathbb{R}^2$, and consider the equation

$$x' = f(t, x) . \tag{4.3}$$

Recall that by a solution of the equation it is understood as any function $x \in C^1(J)$ defined on some interval J of the real line such that $(t, x(t)) \in D$ and $x'(t) = f(t, x(t))$ for every $t \in J$.

Additionally, for a given point $(t_0, x_0) \in D$ (initial data), by the Cauchy problem

$$\begin{cases} x' = f(t, x) \\ x(t_0) = x_0 , \end{cases} \tag{4.4}$$

we mean the problem of finding a solution of the equation (4.3) for which $x(t_0) = x_0$.

We say that the property of uniqueness of solutions holds for the Cauchy problem related to the equation (4.3) if any two solutions of (4.3), which are equal at some point, coincide on the whole intersection of their domains.

The goal of this section is to give a sufficient condition for this uniqueness property to hold. This condition will be given in terms of f.

We say that the function $f(t, x)$ is *locally Lipschitz*[4] *continuous in* x (on D), if for each compact set K included in D, there exists a constant (called a *Lipschitz constant*) L_K such that
$$|f(t, x) - f(t, y)| \leq L_K |x - y|$$
for all $(t, x), (t, y) \in K$. In the case where constant L_K does not depend on K, f is said to be (globally) *Lipschitz continuous in* x. Hence, f is Lipschitz continuous in x if there exists a constant L such that
$$|f(t, x) - f(t, y)| \leq L |x - y|$$
for every $(t, x), (t, y) \in D$.

Theorem 4.5 (Uniqueness). *If f is continuous and locally Lipschitz continuous in x, then the uniqueness of solutions holds for the Cauchy problem related to the equation* (4.3).

Proof. Let $x_1 \in C^1(J_1)$ and $x_2 \in C^1(J_2)$ be two solutions of (4.3) that are equal at some point t_0. We want to prove that
$$x_1(\tau) = x_2(\tau)$$
for all $\tau \in J_1 \cap J_2$. To this aim, let $\tau \in J_1 \cap J_2$ be arbitrary fixed. Assume first that $\tau > t_0$. Then $[t_0, \tau] \subset J_1 \cap J_2$, and since x_1, x_2 are solutions, one has
$$x_1(t) = x_0 + \int_{t_0}^{t} f(s, x_1(s)) \, ds,$$
$$x_2(t) = x_0 + \int_{t_0}^{t} f(s, x_2(s)) \, ds,$$
for all $t \in [t_0, \tau]$, where $x_0 = x_1(t_0) = x_2(t_0)$. It follows that
$$|x_1(t) - x_2(t)| = \left| \int_{t_0}^{t} f(s, x_1(s)) \, ds - \int_{t_0}^{t} f(s, x_2(s)) \, ds \right|$$
$$= \left| \int_{t_0}^{t} (f(s, x_1(s)) - f(s, x_2(s))) \, ds \right|$$
$$\leq \int_{t_0}^{t} |f(s, x_1(s)) - f(s, x_2(s))| \, ds,$$
for all $t \in [t_0, \tau]$. The functions x_1 and x_2 being continuous, their graphs restricted to the compact interval $[t_0, \tau]$ are compact sets in $\mathbb{R}^2$. Let K be the union of these two

---

[4] Rudolf Otto Sigismund Lipschitz (1832–1903)

graphs. Then, in view of our hypothesis, for every $s \in [t_0, \tau]$,

$$|f(s, x_1(s)) - f(s, x_2(s))| \le L_K |x_1(s) - x_2(s)|,$$

and consequently

$$|x_1(t) - x_2(t)| \le L_K \int_{t_0}^{t} |x_1(s) - x_2(s)| \, ds,$$

for all $t \in [t_0, \tau]$. This, according to Corollary 4.4, implies $|x_1(t) - x_2(t)| = 0$ for every $t \in [t_0, \tau]$. Hence $x_1(t) = x_2(t)$ for every $t \in [t_0, \tau]$, and in particular $x_1(\tau) = x_2(\tau)$, as desired. The case when $\tau < t_0$ can be discussed analogously. □

Remark 4.6. If the equation (4.3) is autonomous, i.e. f does not depend on t, and $f \in C^1(D_0)$, where D_0 is an interval of the real line, then f is continuous and locally Lipschitz continuous in x on $D = \mathbb{R} \times D_0$. Indeed, if K is a compact set with $K \subset D$, then there exists a closed bounded interval $[a, b] \subset D_0$, such that $K \subset \mathbb{R} \times [a, b]$. Let $(t, x), (t, y) \in K$. Then $x, y \in [a, b]$, and according to the mean value theorem, there exists a point c between x and y, such that

$$|f(x) - f(y)| = \left| f'(c)(x - y) \right|.$$

Since f' is continuous on $[a, b]$ and $c \in [a, b]$, one has $|f'(c)| \le L_K$, where $L_K = \max_{\tau \in [a,b]} |f'(\tau)|$. Hence

$$|f(x) - f(y)| \le L_K |x - y|.$$

Note that f is even globally Lipschitz continuous in x on D, if f' is bounded on D_0.

Remark 4.7. If the equation in (4.3) is linear, i.e. $f(t, x) = a(t)x + b(t)$, for some $a, b \in C(J)$ and a real interval J, then f is continuous and locally Lipschitz continuous in x on $D := J \times \mathbb{R}$. Indeed, if K is a compact set with $K \subset D$, then there exists a closed bounded interval $[\alpha, \beta] \subset J$ such that $K \subset [\alpha, \beta] \times \mathbb{R}$. Then, for every $(t, x), (t, y) \in K$, one has

$$|f(t, x) - f(t, y)| = |a(t)(x - y)| \le L_K |x - y|,$$

where $L_K = \max_{t \in [\alpha, \beta]} |a(t)|$. Thus, f is locally Lipschitz continuous in x on $D := J \times \mathbb{R}$. Therefore, the uniqueness theorem applies in particular to the linear case.

Note that if the function a is bounded on J, then $f(t, x) = a(t)x + b(t)$ is even globally Lipschitz continuous in x on D.

Example 4.8. Show that if the set $D \subset \mathbb{R}^2$ is open, then the necessary and sufficient condition for a function $f: D \to \mathbb{R}$, $f = f(t, x)$, to be locally Lipschitz continuous in x is such that for each point $(t_0, x_0) \in D$, there is a neighborhood V of (t_0, x_0), $V \subset D$, and a number L_V such that

$$|f(t, x) - f(t, y)| \le L_V |x - y| \tag{4.5}$$

for all $(t, x), (t, y) \in V$.

Solution. The necessity is immediate since for each point of D we may take as neighborhood V a closed ball centered at that point and included in D. Such a set is compact, so the existence of a Lipschitz constant satisfying (4.5) is guaranteed by the definition of the local Lipschitz continuity in x. To prove that the condition is sufficient, take any compact set K included in D. Then, according to the above condition, for each of the points of K there is a neighborhood, now assumed to be open, on which a Lipschitz condition holds. All these neighborhoods represent an open cover of the compact K. Thus there is a finite subcover of K, and we may take as L_K the maximum of the Lipschitz constants corresponding to the component sets of the subcover.

Example 4.9. If $D \subset \mathbb{R}^2$ is open and for a function $f : D \to \mathbb{R}$, and the partial derivative $\partial f(t, x)/\partial x$ exists and is continuous on D, then f is locally Lipschitz continuous in x.

Solution. We use the equivalent condition given in the previous example. Let (t_0, x_0) be an arbitrary point of the open set D. Then, there is a neighborhood V of (t_0, x_0) included in D, of the form $V = [t_0-h, t_0+h] \times [x_0-h, x_0+h]$. For any points $(t, x), (t, y) \in V$, by virtue of the mean value theorem, there is $z \in [x_0 - h, x_0 + h]$ with

$$|f(t, x) - f(t, y)| = \left|\frac{\partial f(t, z)}{\partial x}\right| |x - y|.$$

Hence (4.5) holds with $L_V = \max_{(t,x) \in V} |\partial f(t, x)/\partial x|$.

Example 4.10. Show that the Cauchy problem $x' = 1 + t + x^2$, $x(0) = 1$ has at most one solution $x \in C^1[0, 1]$.

Solution. The function $f : D \to \mathbb{R}$, $f(t, x) = 1 + t + x^2$, where $D = [0, 1] \times \mathbb{R}$ is continuous and locally Lipschitz continuous in x. Thus Theorem 4.5 gives the conclusion.

Example 4.11. Verify that the Cauchy problem $x' = \sqrt{x}$, $x(0) = 0$ admits the solutions $x_1 = 0$ and $x_2 = t^2/4$ for $t \geq 0$. Does this example contradict the uniqueness theorem?

Solution. The function $f(t, x) = \sqrt{x}$ is continuous on $D = \mathbb{R}^2_+$, but is not locally Lipschitz continuous in x. To see this, take the compact $K = [0, 1]^2 \subset D$. If there exists a constant L such that $|\sqrt{x} - \sqrt{y}| \leq L|x - y|$ for every $(t, x), (t, y) \in K$, then $1 \leq L|\sqrt{x} + \sqrt{y}|$ for every $x, y \in [0, 1]$, $x \neq y$, which is impossible. Hence this example does not contradict the uniqueness theorem.

Example 4.12. Show that the Cauchy problem $x' = \sqrt{x}$, $x(0) = 1$ has a unique solution for $t \geq 0$, namely $x = (t/2 + 1)^2$. What does this example say in view of the fact that the function $f(t, x) = \sqrt{x}$ is not locally Lipschitz continuous in x on $\mathbb{R}^2_+$?

Solution. From $x' = \sqrt{x} \geq 0$, we deduce that all solutions are nondecreasing. Then, if x is solution of the given Cauchy problem, in view of $x(0) = 1$, we have $x(t) \geq 1$ for all $t \geq 0$. Thus we may divide into the equation by $\sqrt{x}$ without any problem, separate variables, and finally find $x = (t/2 + 1)^2$. This example shows that the assumption of Theorem 4.5 gives only a sufficient condition, not also a necessary one. Note however that Theorem 4.5 applies if one takes $D = [0, \infty) \times (0, \infty)$.

4.4 Continuous Dependence of Solutions on the Initial Values

The function f being continuous and locally Lipschitz continuous in x, not only guarantees the uniqueness of solutions for the Cauchy problem, but also their continuous dependence on initial values.

Consider again the equation

$$x' = f(t, x), \tag{4.6}$$

and assume that $f: D \subset \mathbb{R}^2 \to \mathbb{R}$ is continuous and locally Lipschitz continuous in x.

First we give another application of Gronwall's inequality, to estimate the distance between two solutions of the equation.

Lemma 4.13. *Let $\varphi \in C^1[a, b]$ be a solution of the equation (4.6), and let $K \subset D$ be a compact set that includes the graph of φ, i.e. $(t, \varphi(t)) \in K$ for every $t \in [a, b]$. Then for any solution $\psi \in C^1[a, b]$ of the equation (4.6) whose graph is completely included in K, and any point $t_0 \in [a, b]$, one has*

$$|\varphi(t) - \psi(t)| \leq |\varphi(t_0) - \psi(t_0)| e^{L_K|t-t_0|}, \quad t \in [a, b]. \tag{4.7}$$

Proof. Since φ and ψ solve the equation (4.6) on the interval $[a, b]$, we have

$$\varphi(t) = \varphi(t_0) + \int_{t_0}^{t} f(s, \varphi(s))\, ds,$$

$$\psi(t) = \psi(t_0) + \int_{t_0}^{t} f(s, \psi(s))\, ds$$

for every $t \in [a, b]$. It follows that

$$|\varphi(t) - \psi(t)| = \left| \varphi(t_0) - \psi(t_0) + \int_{t_0}^{t} (f(s, \varphi(s)) - f(s, \psi(s)))\, ds \right|$$

$$\leq |\varphi(t_0) - \psi(t_0)| + \left| \int_{t_0}^{t} |f(s, \varphi(s)) - f(s, \psi(s))|\, ds \right|.$$

Note that the absolute value of the last integral is taken in order to treat the cases $t_0 < t$ and $t < t_0$ simultaneously. Since the graphs of φ and ψ are completely included in K, we may use the Lipschitz condition in the last integral, and thus

$$|\varphi(t) - \psi(t)| \leq |\varphi(t_0) - \psi(t_0)| + L_K \left| \int_{t_0}^{t} |\varphi(s) - \psi(s)|\, ds \right|.$$

The conclusion now follows from Gronwall's inequality. □

This inequality gives an estimation of the distance between the two solutions, in terms of the distance between their values at a given point $t = t_0$. It shows that, on a given bounded interval, the distance between the solutions is as small as we wish, provided that the distance between their values at any point of that interval is small enough.

The next result concerns the continuous dependence of the solution of the Cauchy problem on the initial value x_0.

Theorem 4.14 (Continuous dependence). *Assume that $\varphi \in C^1[a, b]$ solves the Cauchy problem (4.4), and that for some $\eta > 0$,*

$$K := \{(t, z) : t \in [a, b], |\varphi(t) - z| \leq \eta\} \subset D.$$

Then, for each $\varepsilon > 0$, there exists a $\delta_\varepsilon > 0$ such that whenever

$$|x_0 - \overline{x}_0| \leq \delta_\varepsilon$$

and $\psi \in C^1[a, b]$ solves the Cauchy problem $x' = f(t, x)$, $x(t_0) = \overline{x}_0$, one has

$$|\varphi(t) - \psi(t)| \leq \varepsilon, \quad t \in [a, b].$$

Proof. First we show that if $|x_0 - \overline{x}_0| \leq \delta_0 := \eta \exp(-L_K(b - a))$, then the graph of ψ is completely included in K. To this aim, consider

$$t_1 := \inf\{t \in [a, t_0] : |\varphi(s) - \psi(s)| \leq \eta \text{ for all } s \in [t, t_0]\},$$
$$t_2 := \sup\{t \in [t_0, b] : |\varphi(s) - \psi(s)| \leq \eta \text{ for all } s \in [t_0, t]\}.$$

Clearly $a \leq t_1 < t_0 < t_2 \leq b$ and

$$|\varphi(s) - \psi(s)| \leq \eta \quad \text{for all } s \in [t_1, t_2]. \tag{4.8}$$

Thus we may apply Lemma 4.13 to the interval $[t_1, t_2]$ instead of $[a, b]$, and obtain

$$\begin{aligned}|\varphi(t) - \psi(t)| &\leq |x_0 - \overline{x}_0| e^{L_K|t - t_0|} \\ &\leq |x_0 - \overline{x}_0| e^{L_K(t_2 - t_1)} \\ &\leq \eta e^{-L_K(b-a)} e^{L_K(t_2 - t_1)}, \quad t \in [t_1, t_2].\end{aligned} \tag{4.9}$$

If $t_1 > a$, then from the definition of t_1, we must have $|\varphi(t_1) - \psi(t_1)| = \eta$. Then $t_2 - t_1 < b - a$, and from (4.9), we deduce

$$\eta = |\varphi(t_1) - \psi(t_1)| \leq \eta e^{-L_K(b-a)} e^{L_K(t_2 - t_1)} < \eta,$$

a contradiction. Hence $t_1 = a$. Similarly, $t_2 = b$. Then, with $[t_1, t_2] = [a, b]$, relation (4.8) shows that the graph of ψ is completely included in K.

Thus, if $|x_0 - \overline{x}_0| \leq \delta_0$, then according to Lemma 4.13,

$$|\varphi(t) - \psi(t)| \leq |x_0 - \overline{x}_0| e^{L_K|t - t_0|}, \quad t \in [a, b].$$

Let $\delta_\varepsilon = \min\{\delta_0, \varepsilon \exp(-L_K(b-a))\}$, and assume that $|x_0 - \bar{x}_0| \leq \delta_\varepsilon$. Then for every $t \in [a, b]$,

$$|\varphi(t) - \psi(t)| \leq |x_0 - \bar{x}_0| e^{L_K|t-t_0|} \leq \delta_\varepsilon e^{L_K(b-a)} \leq \varepsilon,$$

as desired. □

Example 4.15. Estimate the distance between two solutions $\varphi, \psi \in C^1[0, a]$ of the equation $x' = t^2 + 2 \sin 3x$, in terms of their initial values $\varphi(0)$ and $\psi(0)$.

Solution. The function $f(t, x) = t^2 + 2 \sin 3x$ is continuous on $D = [0, a] \times \mathbb{R}$ and Lipschitz continuous in x; more exactly, $|f(t, x) - f(t, y)| \leq 6|x - y|$ for all $t \in [0, a]$ and $x, y \in \mathbb{R}$. Using (4.7) we obtain the estimate $|\varphi(t) - \psi(t)| \leq |\varphi(0) - \psi(0)| \exp(6a)$, $t \in [0, a]$.

4.5 The Cauchy Problem for Systems

The results presented in the previous sections can be easily extended from equations to n-dimensional systems by a simple replacement of the absolute value $|\cdot|$ with the Euclidean norm $\|\cdot\|$ of $\mathbb{R}^n$.

Consider the n-dimensional system

$$\begin{cases} x'_1 = f_1(t, x_1, x_2, \ldots, x_n) \\ x'_2 = f_2(t, x_1, x_2, \ldots, x_n) \\ \ldots \\ x'_n = f_n(t, x_1, x_2, \ldots, x_n), \end{cases} \quad (4.10)$$

subject to the initial conditions

$$x_1(t_0) = x_1^0, \quad x_2(t_0) = x_2^0, \quad \ldots, \quad x_n(t_0) = x_n^0, \quad (4.11)$$

where the functions f_i are defined on a set $D \subset \mathbb{R}^{n+1}$, and $(t_0, x_1^0, x_2^0, \ldots, x_n^0) \in D$. Using the notations

$$x = \begin{bmatrix} x_1 \\ x_2 \\ \vdots \\ x_n \end{bmatrix}, \quad x_0 = \begin{bmatrix} x_1^0 \\ x_2^0 \\ \vdots \\ x_n^0 \end{bmatrix}, \quad f(t, x) = \begin{bmatrix} f_1(t, x_1, x_2, \ldots, x_n) \\ f_2(t, x_1, x_2, \ldots, x_n) \\ \vdots \\ f_n(t, x_1, x_2, \ldots, x_n) \end{bmatrix},$$

the system (4.10) can be represented by a single vector equation,

$$x' = f(t, x),$$

where it is understood that f is a vector-valued function from $D \subset \mathbb{R}^{n+1}$ to $\mathbb{R}^n$. The initial conditions (4.11) are also represented by a single vector equality,

$$x(t_0) = x_0,$$

where $x_0 \in \mathbb{R}^n$.

For vector-valued functions, the notions of local and global Lipschitz continuity are defined accordingly. Thus, we say that the vector-valued function $f(t, x)$ is *locally Lipschitz continuous in x*, if for each compact set K included in D, there exists a constant L_K such that

$$\|f(t, x) - f(t, y)\| \le L_K \|x - y\|$$

for all $(t, x), (t, y) \in K$, where in this case the Euclidian norm $\|\cdot\|$ is used instead of the absolute value.

It is worth noting that the property of a vector-valued function f of being locally Lipschitz continuous in x can be equivalently defined by means of separate Lipschitz conditions for the scalar components f_i of the function. Thus f is locally Lipschitz continuous in x, if for each compact set $K \subset D$, there exist constants $l_{ij}^K \in \mathbb{R}_+$ for $i, j = 1, 2, \ldots, n$, such that

$$|f_i(t, x_1, x_2, \ldots, x_n) - f_i(t, y_1, y_2, \ldots, y_n)|$$
$$\le l_{i1}^K |x_1 - y_1| + l_{i2}^K |x_2 - y_2| + \ldots + l_{in}^K |x_n - y_n|, \quad i = 1, 2, \ldots, n,$$

for every $(t, x_1, x_2, \ldots, x_n), (t, y_1, y_2, \ldots, y_n) \in K$. The proof of the equivalence of the two definitions is left to the reader as an exercise.

Note again that in the linear case, when the functions f_i are of the form

$$f_i(t, x_1, x_2, \ldots, x_n) = a_{i1}(t)x_1 + a_{i2}(t)x_2 + \ldots + a_{in}(t)x_n + b_i(t)$$

with $a_{ij}, b_i \in C(J)$, we may assert that f is continuous and locally Lipschitz continuous in x on $D = J \times \mathbb{R}^n$. If, in addition, the functions a_{ij} are bounded on the interval J, then f is globally Lipschitz continuous in x.

Taking into consideration the above remarks, the results on uniqueness and continuous dependence of solutions on the initial values remain true for n-dimensional differential systems. In particular, the estimation formula (4.7) reads for systems as

$$\|\varphi(t) - \psi(t)\| \le \|\varphi(t_0) - \psi(t_0)\| e^{L_K|t-t_0|}, \quad t \in [a, b].$$

4.6 The Cauchy Problem for Higher-Order Equations

Consider the differential equation of order n

$$x^{(n)} = g\left(t, x, x', \ldots, x^{(n-1)}\right), \tag{4.12}$$

subject to the initial conditions

$$x(t_0) = x_1^0, \quad x'(t_0) = x_2^0, \quad \ldots, \quad x^{(n-1)}(t_0) = x_n^0, \tag{4.13}$$

where $g: D \subset \mathbb{R}^{n+1} \to \mathbb{R}$ is a continuous function, $(t_0, x_0) \in D$ and $x_0 = (x_1^0, x_2^0, \ldots, x_n^0)$.

As already known, by making the notations

$$x_1 := x, \quad x_2 := x', \quad \ldots, \quad x_n := x^{(n-1)},$$

the equation is reduced to the system

$$\begin{cases} x_1' = x_2 \\ x_2' = x_3 \\ \ldots \\ x_{n-1}' = x_n \\ x_n' = g(t, x_1, x_2, \ldots, x_n), \end{cases} \quad (4.14)$$

and the initial conditions read as

$$x_1(t_0) = x_1^0, \quad x_2(t_0) = x_2^0, \quad \ldots, \quad x_n(t_0) = x_n^0.$$

Let us remark that the right-hand sides of the first $n-1$ equations of the system (4.14) are linear. Thus in order to apply the previous considerations on the Cauchy problem to the particular system (4.14), we need $g(t, x)$ to be locally Lipschitz continuous in x. For such a function, this means that for each compact $K \subset D$, there exist constants $l_j^K \in \mathbb{R}_+, j = 1, 2, \ldots, n$, such that

$$|g(t, x_1, x_2, \ldots, x_n) - g(t, y_1, y_2, \ldots, y_n)|$$
$$\leq l_1^K |x_1 - y_1| + l_2^K |x_2 - y_2| + \ldots + l_n^K |x_n - y_n|$$

for every $(t, x_1, x_2, \ldots, x_n), (t, y_1, y_2, \ldots, y_n) \in K$.

4.7 Periodic Solutions

An interesting consequence of the property of uniqueness of solutions for the Cauchy problem concerns *periodic solutions* of autonomous differential systems.

Theorem 4.16. *Assume that the uniqueness of solutions holds for the Cauchy problem related to the autonomous system $x' = f(x)$, where $f: \mathbb{R}^n \to \mathbb{R}^n$. If for a solution $x \in C^1(\mathbb{R}, \mathbb{R}^n)$ of the autonomous system there exist two distinct numbers a and b ($a < b$) such that*

$$x(a) = x(b), \quad (4.15)$$

then the solution x is periodic and the number $b - a$ is a period of x.

Proof. One can easily check that the function

$$y(t) := x(t + b - a)$$

is also a solution of the differential system. In addition, $y(a) = x(b)$. Then from (4.15), $y(a) = x(a)$ and so x and y are solutions of the same Cauchy problem. Now the uniqueness of solutions for the Cauchy problem for systems guarantees that $x = y$, that is

$$x(t) = x(t + b - a) \quad \text{for every} \quad t \in \mathbb{R}.$$

Hence x is a periodic function and $b - a$ is one of its periods (not necessarily the principal period). □

Remark 4.17. Theorem 4.16 implies that the closed orbits of any autonomous system for which the uniqueness of solutions for the Cauchy problem holds, correspond to the periodic solutions of the system.

The analogue of Theorem 4.16 for the nonautonomous case is contained in the following example:

Example 4.18. Assume that the uniqueness of solutions for the Cauchy problem holds for the system $x' = f(t, x)$, where $f : \mathbb{R} \times \mathbb{R}^n \to \mathbb{R}^n$, and that $f(\cdot, x)$ is periodic with period p for every $x \in \mathbb{R}^n$. If for a solution $x \in C^1(\mathbb{R}, \mathbb{R}^n)$ of the system $x' = f(t, x)$ there exists a number a such that

$$x(a) = x(a + p), \tag{4.16}$$

then the solution x is periodic with period p.

Solution. Show that the function $y(t) := x(t + p)$ is also a solution. Since $y(a) = x(a + p) = x(a)$, the two functions x and y solve the same Cauchy problem. Hence $x = y$, that is $x(t) = x(t + p)$ for every $t \in \mathbb{R}$.

For first-order differential equations on the real line, there is a necessary and sufficient condition for the existence of periodic solutions, namely Massera's[5] theorem.

Theorem 4.19 (Massera's theorem). *Assume that the uniqueness of solutions holds for the Cauchy problem related to the equation $x' = f(t, x)$, where $f : \mathbb{R} \times \mathbb{R} \to \mathbb{R}$ is continuous, and that $f(\cdot, x)$ is periodic with period p for every $x \in \mathbb{R}$. Then the equation $x' = f(t, x)$ has a periodic solution of period p if and only if it has a bounded solution on the real line.*

Proof. The necessity of the condition is obvious since any periodic solution is bounded on the real line. It remains to show that if there is a bounded solution $x_0 \in C^1(\mathbb{R})$, then there is also a periodic solution of period p. The proof is constructive. For every integer $k \geq 0$, we define the function x_k by

$$x_k(t) := x_0(t + kp).$$

Due to the periodicity of f in t, every function x_k is a solution of the given equation.

---

[5] José Luis Massera (1915–2002)

If there is t_0 such that $x_1(t_0) = x_0(t_0)$, then based on the uniqueness property, $x_1 = x_0$, that is $x_0(t + p) = x_0(t)$ for all $t \in \mathbb{R}$, which means that x_0 is a periodic solution of period p, and the proof is finished.

Assume that $x_1(t) \neq x_0(t)$ for every t. Then either $x_1 > x_0$ on $\mathbb{R}$, or $x_1 < x_0$ on $\mathbb{R}$. Let us assume that $x_1 > x_0$ on $\mathbb{R}$. Then $x_1(t + kp) > x_0(t + kp)$, i.e.

$$x_{k+1}(t) > x_k(t) \quad \text{for every } t \in \mathbb{R}.$$

Thus the sequence of functions (x_k) is increasing. It is also uniformly bounded in virtue of the boundedness of x_0. Hence there is a function x_∞ such that

$$\lim_{k \to \infty} x_k(t) = x_\infty(t) \quad \text{(pointwise)}.$$

From $x'_k = f(t, x_k)$, it follows that the derivatives of x_k are also uniformly bounded. Consequently, the sequence (x_k) is equicontinuous. Then $\lim_{k \to \infty} x_k = x_\infty$ uniformly on any compact interval of the real line. Then we can pass to the limit in

$$x_k(t) = x_k(0) + \int_0^t f(s, x_k(s))\, ds,$$

to obtain

$$x_\infty(t) = x_\infty(0) + \int_0^t f(s, x_\infty(s))\, ds.$$

This shows that x_∞ is a solution of the given equation. On the other hand,

$$x_\infty(t) = \lim_{k \to \infty} x_k(t) = \lim_{k \to \infty} x_{k+1}(t) = \lim_{k \to \infty} x_k(t + p) = x_\infty(t + p).$$

Hence x_∞ is periodic with period p. The case $x_1 < x_0$ can be treated in a similar way. □

Example 4.20. If $f : \mathbb{R}^n \to \mathbb{R}^n$ is continuous and $f(x) \cdot x \neq 0$ for every $x \in \mathbb{R}^n \setminus \{0\}$, where $\cdot$ stands for the scalar product in $\mathbb{R}^n$, then the autonomous system $x' = f(x)$ has no periodic nonzero solutions.

Solution. For an arbitrary nonzero solution $x \in C^1(\mathbb{R}, \mathbb{R}^n)$ of the system, from $x'(t) = f(x(t))$, taking the scalar product with $x(t)$ yields

$$\frac{1}{2}\left(\|x(t)\|^2\right)' = f(x(t)) \cdot x(t), \quad t \in \mathbb{R}.$$

On each interval where $x \neq 0$, we have either $f(x(t)) \cdot x(t) > 0$, and the function $\|x(t)\|^2$ is strictly increasing, or $f(x(t)) \cdot x(t) < 0$, and the function $\|x(t)\|^2$ is strictly decreasing. However this is impossible if $x(t)$ is periodic, when $\|x(t)\|^2$ is also periodic.

Notice that for planar systems (when $n = 2$), the weaker assumption $f(x) \neq 0$ for all $x \in \mathbb{R}^2$ is sufficient for the system $x' = f(x)$ to have no periodic solutions [9, p. 252], [18]. The proof of this result, which gives us the so-called *equilibrium criterion*, is much more complex. Other criteria for existence or nonexistence of periodic solutions of planar systems may be found in the Bibliography.

Example 4.21. The system $x_1' = 2x_1 - x_2 + x_1^3$, $x_2' = x_1 + 3x_2$ has no periodic nonzero solutions.

Solution. We have $f(x) = f(x_1, x_2) = (2x_1 - x_2 + x_1^3, x_1 + 3x_2)$. Then

$$f(x) \cdot x = (2x_1 - x_2 + x_1^3)x_1 + (x_1 + 3x_2)x_2$$
$$= 2x_1^2 + x_1^4 + 3x_2^2 > 0$$

for every $x \neq 0$. Thus we can apply the result from the previous example.

Example 4.22. The system $x' = x^2 + 3y^2$, $y' = 1 + xy$ has no periodic solutions.

Solution. The algebraic system $x^2 + 3y^2 = 0$, $1 + xy = 0$ has no solutions. Hence the differential system does not have any equilibria, and according to the equilibrium criterion, it has no periodic solutions.

4.8 Picard's Method of Successive Approximations

So far, the existence of solutions for the Cauchy problem has been clarified only for linear differential systems with constant coefficients. Now we are going to present one of the most commonly used methods for the treatment of nonlinear equations, namely, *Picard's*[6] *method of successive approximations*, or the method of *Picard's iteration*. The method is used to prove the existence of solutions, and also for finding approximate solutions. For simplicity, we shall present the method for the Cauchy problem related to a first-order differential equation. The technique can be easily extended to n-dimensional systems by replacing the absolute value $|\cdot|$ with the Euclidian norm $\|\cdot\|$ of $\mathbb{R}^n$.

4.8.1 Picard's Iteration

Consider the Cauchy problem

$$\begin{cases} x' = f(t, x), & t \in J \\ x(t_0) = \alpha. \end{cases} \quad (4.17)$$

---

6 Charles Émile Picard (1856–1941)

Assume that $f: J \times \mathbb{R} \to \mathbb{R}$ is a continuous function, where J is an interval of real numbers containing t_0. As we already know, the problem is equivalent to the integral equation

$$x(t) = \alpha + \int_{t_0}^{t} f(s, x(s))\, ds, \quad t \in J. \tag{4.18}$$

The method of successive approximations starts with the choice of a continuous function $x_0 \in C(J)$, usually the constant function $x_0(t) \equiv \alpha$, which is considered a first approximation of the unknown solution of the equation (4.18). Next, the second approximation of the solution is considered to be

$$x_1(t) = \alpha + \int_{t_0}^{t} f(s, x_0(s))\, ds, \quad t \in J.$$

Continuing in this way, the $(k+1)$st approximation x_{k+1} is defined using x_k by

$$x_{k+1}(t) = \alpha + \int_{t_0}^{t} f(s, x_k(s))\, ds, \quad t \in J. \tag{4.19}$$

Thus we have arrived at the so-called *sequence of successive approximations* (x_k).

In what follows, we consider that the first approximation is the constant function α, i.e. $x_0(t) \equiv \alpha$. The connection between the sequence of successive approximations and the Cauchy problem itself is given by the following result:

Theorem 4.23. *If there exists a function $x \in C(J)$ such that the sequence $(x_k(t))$ converges to $x(t)$ uniformly on every compact interval included in J, then x is a solution of the Cauchy problem.*

Proof. Under the assumptions of the theorem, we may pass to the limit in (4.19) with $k \to \infty$ and we obtain

$$x(t) = \alpha + \int_{t_0}^{t} f(s, x(s))\, ds, \quad t \in J.$$

This shows that x is a solution of the Cauchy problem. □

Example 4.24. Define the sequence of successive approximations for the Cauchy problem $x' = x$, $x(0) = 1$. Find the exact solution and observe its connection with the sequence of successive approximations.

Solution. The equivalent integral equation is

$$x(t) = 1 + \int_0^t x(s)\, ds\,.$$

As a first approximation of the solution we take $x_0(t) \equiv 1$. Then

$$x_1(t) = 1 + \int_0^t 1 \, ds = 1 + t,$$

$$x_2(t) = 1 + \int_0^t (1+s) \, ds = 1 + t + \frac{1}{2}t^2,$$

$$x_3(t) = 1 + \int_0^t \left(1 + s + \frac{1}{2}s^2\right) ds = 1 + t + \frac{1}{2}t^2 + \frac{1}{6}t^3.$$

Using mathematical induction yields the general term of the sequence,

$$x_k(t) = \sum_{i=0}^{k} \frac{1}{i!} t^i.$$

The limit of this sequence is e^t and so e^t is the (unique) solution of the Cauchy problem. Solving the separable equation also gives the exact solution.

In most cases, contrary to the previous example, one cannot explicitly give the sequence of successive approximation, and even less so its limit or the exact solution.

Example 4.25. Write down the recurrence formula defining the sequence of successive approximations for the Cauchy problem $x' = t + x^2$, $x(0) = 1$, and calculate the first three approximations.

Solution. The equivalent integral equation is

$$x(t) = 1 + \int_0^t (s + x(s)^2) \, ds,$$

or

$$x(t) = 1 + \frac{t^2}{2} + \int_0^t x(s)^2 \, ds.$$

The iterative approximation schema is

$$x_0(t) \equiv 1,$$

$$x_{k+1}(t) = 1 + \frac{t^2}{2} + \int_0^t x_k(s)^2 \, ds, \quad k = 0, 1, 2, \ldots.$$

The first three approximations of the solution are: $x_0(t) \equiv 1$, $x_1(t) = 1 + t^2/2 + t$, and

$$x_2(t) = 1 + \frac{t^2}{2} + \int_0^t \left(1 + \frac{s^2}{2} + s\right)^2 ds$$

$$= 1 + t + \frac{3t^2}{2} + \frac{2t^3}{3} + \frac{t^4}{4} + \frac{t^5}{20}.$$

4.8.2 The Interval of Picard's Iteration

The first question concerning Picard's iteration is on what interval are the iterates well-defined. The answer is simple if f is a continuous function on a set of the form $J \times \mathbb{R}$, as is the case for linear equations; then the successive approximations are defined themselves on the interval J. But, if f is only defined on a subset D of $J \times \mathbb{R}$, then there is a risk that at some iteration k there will exist values of s from J such that $(s, x_k(s)) \notin D$, which makes the definition on the whole interval J of the next approximation x_{k+1} impossible. For such cases, the next result guarantees that the sequence of successive approximations is well defined at least on some neighborhood of t_0.

Theorem 4.26. *Let f be defined and continuous on the set*

$$\Delta := \{(t, x): |t - t_0| \le a, |x - \alpha| \le b\},$$

where $0 < a < +\infty$ and $0 < b \le +\infty$. Then the sequence of successive approximations is well defined on the interval $J_h := [t_0 - h, t_0 + h]$, where $h = \min\{a, b/M\}$ and $M = \max_{(t,x) \in \Delta} |f(t, x)|$, in the case where b is finite, and $h = a$, if $b = +\infty$.

Proof. The case $b = +\infty$ being clear, we discuss only the situation when $b < +\infty$. We have to show that the graph of any function x_k from the sequence of successive approximations (where the initial approximation is $x_0(t) \equiv \alpha$), is entirely included in Δ for $t \in J_h$. We proceed by mathematical induction. The initial step of induction being clear, it remains to prove the implication

$$\{(t, x_k(t)): t \in J_h\} \subset \Delta \quad \text{implies} \quad \{(t, x_{k+1}(t)): t \in J_h\} \subset \Delta.$$

Let $t \in J_h$. One has

$$|x_{k+1}(t) - \alpha| = \left| \int_{t_0}^{t} f(s, x_k(s))\, ds \right| \le \left| \int_{t_0}^{t} |f(s, x_k(s))|\, ds \right|.$$

From the induction hypothesis, we have that $(s, x_k(s)) \in \Delta$ for all $s \in J_h$. Then $|f(s, x_k(s))| \le M$ and hence

$$|x_{k+1}(t) - \alpha| \le M |t - t_0| \le Mh \le b,$$

which proves that $(t, x_{k+1}(t)) \in \Delta$ as desired. □

Remark 4.27. The handy choice $x_0 = \alpha$ for the first approximation is not essential. The theorem is true for any choice of the first approximation x_0 for which $|x_0(t) - \alpha| \le b$ for every $t \in J_h$.

Example 4.28. Give an interval of Picard's iteration for the Cauchy problem $x' = \sqrt{1 - t^2} + \sqrt{4 - x^2}$, $x(0) = 0$.

Solution. The function $f(t, x) = \sqrt{1 - t^2} + \sqrt{4 - x^2}$ is defined on the set $\Delta = [-1, 1] \times [-2, 2]$. Hence $a = 1$ and $b = 2$. It is easy to see that $M = 1 + 2 = 3$, and $h = \min\{1, 2/3\} = 2/3$. Thus the interval of Picard's iteration is $[-2/3, 2/3]$.

Example 4.29. Define the sequence of successive approximations and determine an interval on which they are well defined for the Cauchy problem $x' = \sqrt{t - x}$, $x(1) = 0$.

Solution. Take as a first approximation $x_0(t) = 0$. Then

$$x_{k+1}(t) = \int_1^t \sqrt{s - x_k(s)}\, ds, \quad k = 0, 1, \ldots.$$

The domain of f is $D = \{(t, x) \in \mathbb{R}^2 : t - x \geq 0\}$. In order to apply Theorem 4.26, we choose a rectangle Δ centered at $(1, 0)$, included in D. Such a set is $\Delta = [1 - a, 1 + a] \times [a - 1, 1 - a]$, where $0 < a < 1$ is chosen arbitrarily (the reader is advised to draw a picture). Then $M = \sqrt{2}$, so $h = h_a = \min\{a, (1 - a)/\sqrt{2}\}$. The required interval is $J_{h_a} := [1 - h_a, 1 + h_a]$. Of course we could be interested in the interval J_{h_a} that is as large as possible. To this aim we look for that value of a in $(0, 1)$ for which h_a is maximal. This optimal value is $a = 1/(1 + \sqrt{2})$, for which $h = 1/(1 + \sqrt{2})$.

4.8.3 Convergence of Picard's Iteration

A second question about Picard's iteration is concerned with the convergence of the sequence of successive approximations. In view of Theorem 4.23, a positive answer to this question will clarify the problem of the existence of solutions to the Cauchy problem. The next result gives an answer to this question.

Theorem 4.30. *If the function f is continuous and Lipschitz continuous in x on Δ, then the sequence of successive approximations is uniformly convergent on the interval J_h.*

Proof. From Theorem 4.26, the sequence of functions (x_k) is well defined on the interval J_h.

As the function f is assumed Lipschitz continuous in x on Δ, there is a constant $L > 0$ such that

$$|f(t, x) - f(t, y)| \leq L\,|x - y|$$

for every $(t, x), (t, y) \in \Delta$. Additionally, since x_0 and x_1 are continuous on J_h, there is a constant $N > 0$ such that

$$|x_1(t) - x_0(t)| \leq N, \quad t \in J_h. \tag{4.20}$$

We prove by mathematical induction that

$$|x_{i+1}(t) - x_i(t)| \leq N\frac{(L\,|t - t_0|)^i}{i!}, \quad t \in J_h \tag{4.21}$$

for all natural numbers i. For $i = 0$, (4.21) is exactly (4.20). Assume that (4.21) holds for some i. Then

$$|x_{i+2}(t) - x_{i+1}(t)| \le \left| \int_{t_0}^{t} |f(s, x_{i+1}(s)) - f(s, x_i(s))| \, ds \right|$$

$$\le L \left| \int_{t_0}^{t} |x_{i+1}(s) - x_i(s)| \, ds \right|$$

$$\le L \left| \int_{t_0}^{t} N \frac{(L|s - t_0|)^i}{i!} \, ds \right|$$

$$= N \frac{(L|t - t_0|)^{i+1}}{(i+1)!}.$$

Hence (4.21) also holds for $i + 1$. Therefore the statement (4.21) holds for all natural numbers i. Furthermore, since $t \in J_h$, one has $|t - t_0| \le h$ and consequently

$$|x_{i+1}(t) - x_i(t)| \le N \frac{(Lh)^i}{i!}, \quad t \in J_h \tag{4.22}$$

for $i = 0, 1, \ldots$. Let us observe that the sequence $(x_k(t))$ represents the partial sums of the function series

$$x_0(t) + \sum_{i \ge 0} (x_{i+1}(t) - x_i(t)). \tag{4.23}$$

Thus the uniform convergence on J_h of the sequence $(x_k(t))$ means the uniform convergence of series (4.23), which via (4.22) is compared to the convergent numerical series

$$\sum_{i \ge 0} N \frac{(Lh)^i}{i!} = Ne^{Lh} < \infty.$$

According to the Weierstrass test, the function series (4.23) is uniformly convergent on J_h. Thus the sequence of successive approximations $(x_k(t))$ is uniformly convergent on J_h, as desired. □

4.9 Existence of Solutions for the Cauchy Problem

Theorems 4.23, 4.30 and 4.5 yield the following existence and uniqueness result:

Theorem 4.31 (Local existence and uniqueness). *If the function f is continuous and Lipschitz continuous in x on Δ, then the Cauchy problem (4.17) has a unique solution defined on the interval J_h.*

Next, coming back to the general case of a function f defined on an arbitrary set $D \subset \mathbb{R}^2$, we can state and prove the following global existence and uniqueness result:

Theorem 4.32 (Global existence and uniqueness). *Let $D \subset \mathbb{R}^2$ be an open set and $(t_0, \alpha) \in D$. Assume that $f \colon D \to \mathbb{R}$ is continuous and locally Lipschitz continuous in x. Then the Cauchy problem (4.17) has a unique saturated solution. Moreover, the domain of the solution is an open interval (t_-, t_+) and any of the limit points of the solution graph as $t \searrow t_-$ and $t \nearrow t_+$ are located either at infinity or on the boundary of D.*

Proof. 1. The local existence and uniqueness theorem guarantees the existence of the solution in a neighborhood of t_0. The largest interval on which this function can be extended as a solution is open, since if for example it were be closed to its right extremity, be it t_+, then applying the local existence and uniqueness theorem to the Cauchy problem
$$y' = f(t, y), \quad y(t_+) = x(t_+)$$
in a neighborhood of type Δ of the point $(t_+, x(t_+))$, it would result in the function x being extendable as solution to the right of t_+, which is excluded by the definition of t_+.

2. Let us now discuss the limit points of the solution graph as $t \nearrow t_+$. The case $t \searrow t_-$ can be treated similarly. If $t_+ = +\infty$, then it is clear that any limit point of the graph as $t \nearrow +\infty$ is a point at infinity. Assume that $t_+ < +\infty$. Then a limit point is of the form (t_+, x_+), where $x_+ = \lim_{k \to \infty} x(t_k)$ for some sequence (t_k) with $t_k \nearrow t_+$. If $x_+ = +\infty$ or $x_+ = -\infty$, then again the limit point is at infinity. Let x_+ be finite like t_+. It is clear that $(t_+, x_+) \in \overline{D}$. If the point (t_+, x_+) were in the open set D, then applying the local existence and uniqueness theorem to the Cauchy problem
$$y' = f(t, y), \quad y(t_+) = x_+,$$
in a neighborhood of type Δ of (t_+, x_+), it would follow that the function x is extendable as a solution to the right of t_+, which is impossible. Hence the limit point (t_+, x_+) belongs to $\overline{D} \setminus D$, that is, to the boundary of D. □

Remark 4.33. Under the assumptions of Theorem 4.32, in the case where $D = \mathbb{R}^2$, one of the following situations holds for the saturated solution x of the Cauchy problem:
(a) $t_+ = +\infty$, i.e. the solution is defined on the whole semiline $[t_0, +\infty)$;
(b) $t_+ < +\infty$ and $\lim_{t \nearrow t_+} |x(t)| = +\infty$, i.e. the solution blows up in finite time, to the right of t_0.
A similar remark also holds for t_-.

If $D = J \times \mathbb{R}$, where J is an interval, and f is continuous on D and Lipschitz continuous in x on each subset of D of the form $[a, b] \times \mathbb{R}$, then the Cauchy problem has a unique saturated solution on the whole interval J, and the sequence of successive approximations converges to the solution uniformly on each compact subset of J. This remark holds generally for n-dimensional systems of the form $x' = f(t, x)$, where $f \colon J \times \mathbb{R}^n \to \mathbb{R}^n$ is continuous and Lipschitz continuous in x on each subset of D of the form $[a, b] \times \mathbb{R}^n$.

4.9 Existence of Solutions for the Cauchy Problem

In particular, for **linear systems** of the form

$$x' = A(t)x + b(t),$$

with $A: J \to \mathcal{M}_n(\mathbb{R})$ and $b: J \to \mathbb{R}^n$ continuous (where $\mathcal{M}_n(\mathbb{R})$ is the set of all square matrices of order n), the Cauchy problem with the initial condition $x(t_0) = x_0$, where $t_0 \in J$ and $x_0 \in \mathbb{R}^n$, has a unique saturated solution defined on the whole interval J, which can be obtained by the method of successive approximations. This makes it possible to extend the structure Theorem 2.5 for homogeneous linear systems to the general case of systems with coefficients not necessarily constant. Thus, in order to show that the solution set of the homogeneous system

$$x' = A(t)x$$

is a linear space of dimension n, one considers any initial point $t_0 \in J$ and the saturated solutions $x_1, x_2, \ldots, x_n$ of the Cauchy problems corresponding to the initial conditions

$$x_i(t_0) = e_i,$$

where $e_i = [0, \ldots, 1, \ldots, 0]^T$, with 1 at the ith position. To prove that these n solutions are linearly independent, assume that there are constants $c_1, c_2, \ldots, c_n$ such that the linear combination

$$\sum_{i=1}^n c_i x_i = 0 \quad \text{on } J.$$

Then, in particular, one has

$$0 = \sum_{i=1}^n c_i x_i(t_0) = \sum_{i=1}^n c_i e_i,$$

which in virtue of the linear independence of the vectors e_i yields $c_i = 0$ for all i. Hence the solutions $x_1, x_2, \ldots, x_n$ are linear independent. To show that these solutions generate the solution set of the homogeneous differential system, take any function $\varphi = [\varphi_1, \varphi_2, \ldots, \varphi_n]^T$ with $\varphi' = A(t)\varphi$ on J. Clearly one has the representation

$$\varphi(t_0) = \sum_{i=1}^n \varphi_i(t_0) e_i.$$

Then the linear combination

$$x := \sum_{i=1}^n \varphi_i(t_0) x_i$$

obviously solves the homogeneous system and the initial condition $x(t_0) = \varphi(t_0)$. Consequently, by the uniqueness of the solution for the Cauchy problem, one has $\varphi = x$, that is φ is a linear combination of the solutions $x_1, x_2, \ldots, x_n$. Therefore, the solutions $x_1, x_2, \ldots, x_n$ represent a base of the solution set of the homogeneous system.

Additionally, the concept of a fundamental matrix, formula (2.23) giving the relationship between different fundamental matrices, the representation Theorem 2.9, and the method of variation of parameters, apply generally to linear systems with not necessarily constant coefficients.

Moreover, the above remarks can be transferred to linear higher-order differential equations with not necessarily constant coefficients via the equivalence to first-order differential systems.

Chapter 5 Stability of Solutions

5.1 The Notion of a Stable Solution

In this chapter we consider n-dimensional systems of the form

$$x' = f(t, x), \qquad (5.1)$$

assuming that f is continuous in (t, x) and locally Lipschitz continuous in x on $D := \mathbb{R}_+ \times D_0$, where D_0 is an open subset of $\mathbb{R}^n$. Then the Cauchy problem with the initial condition $x(t_0) = x_0$, where $(t_0, x_0) \in D$, has a unique saturated solution, which is denoted by $x(t, t_0, x_0)$. According to the results from Section 4.4, if a solution φ of the system is defined on some compact interval $[t_0, t_0 + T]$, then it depends continuously on the initial value; more exactly,

for each $\varepsilon > 0$, there is $\delta = \delta(\varepsilon, t_0) > 0$ such that
$$\|x_0 - \varphi(t_0)\| \leq \delta \text{ implies} \qquad (5.2)$$
$$\|x(t, t_0, x_0) - \varphi(t)\| \leq \varepsilon \text{ for all } t \in [t_0, t_0 + T].$$

The situation may be different if the compact interval $[t_0, t_0 + T]$ is replaced by the unbounded interval $[t_0, +\infty)$. To see this, it suffices to consider the equation $x' = x$, for which $x(t, t_0, x_0) = x_0 e^{t-t_0}$, and to observe that for any distinct values x_0, y_0, no matter how close, one has

$$|x(t, t_0, y_0) - x(t, t_0, x_0)| = |y_0 - x_0| e^{t-t_0} \to +\infty \text{ as } t \to +\infty.$$

If for a solution $\varphi \in C^1(\mathbb{R}_+)$ of the system, a similar property to (5.2) holds on the whole semiline $[t_0, +\infty)$, then the solution φ is said to be stable. More precisely, we have the following definitions:

Definition 5.1. A solution $\varphi \in C^1(\mathbb{R}_+, \mathbb{R}^n)$ of the system (5.1) is said to be *stable* (in the sense of Lyapunov[1]) if

for each $\varepsilon > 0$ and $t_0 \geq 0$, there is $\delta = \delta(\varepsilon, t_0) > 0$ such that
$$\|x_0 - \varphi(t_0)\| \leq \delta \text{ implies} \qquad (5.3)$$
$$\|x(t, t_0, x_0) - \varphi(t)\| \leq \varepsilon \text{ for all } t \in [t_0, +\infty).$$

A solution that is not stable is said to be *unstable*.

Definition 5.2. A solution $\varphi \in C^1(\mathbb{R}_+, \mathbb{R}^n)$ of the system (5.1) is said to be *asymptotically stable* if it is stable and for each $t_0 \geq 0$, there exists $\eta = \eta(t_0) > 0$ such that

$$\|x_0 - \varphi(t_0)\| \leq \eta \quad \text{implies} \quad \lim_{t \to +\infty} \|x(t, t_0, x_0) - \varphi(t)\| = 0.$$

---

[1] Aleksandr Mikhailovich Lyapunov (1857–1918)

If the number δ in Definition 5.1 can be found independent of t_0, i.e. $\delta = \delta(\varepsilon)$, then the solution φ is said to be *uniformly stable*. A uniformly stable solution that has the additional property from Definition 5.2, with the number η not depending on t_0, is said to be *uniformly asymptotically stable*.

Example 5.3. Consider the equation $x' = \lambda x$, where λ is a real number. In this case, $x(t, t_0, x_0) = x_0 \exp(\lambda(t - t_0))$. For $\lambda < 0$, any solution is asymptotically stable; for $\lambda = 0$, all solutions are constant and are stable, but not asymptotically stable; for $\lambda > 0$, all solutions are unstable.

Example 5.4. Consider the separable equation $x' = x - x^2$. One has $x(t, t_0, x_0) = [1 - (1 - 1/x_0)\exp(t_0 - t)]^{-1}$ for $x_0 \neq 0$, and $x(t, t_0, 0) = 0$. Since $x(t, t_0, x_0) \to 1$ as $t \to +\infty$ ($x_0 \neq 0$), the zero solution is unstable, while the solution $x = 1$ is asymptotically stable.

It is important to note that the stability investigation of any solution φ of the system (5.1) can be reduced to the stability investigation of the zero solution of an associated system. Indeed, if we make the change of variable $y := x - \varphi$, then the system (5.1) in x becomes a system in y, namely

$$y' = f(t, y + \varphi) - \varphi'(t), \qquad (5.4)$$

and to the solution $x = \varphi$ of (5.1) corresponds the solution $y = 0$ of (5.4). In addition, it is easy to check that φ is stable (asymptotically stable) for the system (5.1) if and only if the zero solution of (5.4) is stable (asymptotically stable, respectively). Thus, without loss of generality, we shall discuss the stability of the zero solution of the system (5.1), under the assumption that $0 \in D_0$ and

$$f(t, 0) = 0, \quad t \in \mathbb{R}_+ .$$

Also note that in the case of autonomous systems, stable equilibrium points and asymptotically stable equilibrium points are necessarily uniformly stable and uniformly asymptotically stable, respectively. This happens due to the obvious formula

$$x(t, t_0, x_0) = x(t - t_0, 0, x_0), \quad t \geq t_0 ,$$

which shows that the graph of the solution starting from (t_0, x_0) is obtained by the translation to the right of the graph of the solution starting from $(0, x_0)$. Thus, in order to study the stability of equilibrium points of autonomous systems, it suffices to discuss only the situation $t_0 = 0$, and therefore to simply denote

$$x(t, x_0) := x(t, 0, x_0) .$$

5.2 Stability of Linear Systems

If the system (5.1) is linear, i.e. it has the form

$$x' = A(t)x + b(t), \qquad (5.5)$$

where $A \in C(\mathbb{R}_+, \mathcal{M}_n(\mathbb{R}))$ and $b \in C(\mathbb{R}_+, \mathbb{R}^n)$, then for each solution $\varphi \in C(\mathbb{R}_+, \mathbb{R}^n)$ of the system, the substitution $y = x - \varphi$ leads to the system (5.4), which is exactly the homogeneous system of (5.5) (the same system for every solution φ),

$$y' = A(t)y.$$

Thus if the zero solution of the homogeneous system is stable or asymptotically stable, then so are all the solutions of the nonhomogeneous system, and vice versa. Therefore, for linear systems, stability is a property of the system, contrary to the case of nonlinear systems, where some of solutions can be stable while others unstable. Hence, we speak about stable solutions in the case of nonlinear systems, and about stable systems in the linear case. Thus, a linear system is stable, or asymptotically stable, if all its solutions are stable, or asymptotically stable respectively.

Theorem 5.5 (Stability of linear systems). (I) *The following statements are equivalent:*
  (a) *The system (5.5) is stable;*
  (b) *The system (5.5) admits a fundamental matrix $U(t)$ bounded on $\mathbb{R}_+$;*
  (c) *Any fundamental matrix of the system (5.5) is bounded on $\mathbb{R}_+$;*
  (d) *All solutions of the associated homogeneous system are bounded on $\mathbb{R}_+$.*
(II) *The following statements are equivalent:*
  (a) *The system (5.5) is asymptotically stable;*
  (b) *The system (5.5) admits a fundamental matrix $U(t)$, which tends to the zero matrix as $t \to +\infty$;*
  (c) *Any fundamental matrix of the system (5.5) tends to the zero matrix as $t \to +\infty$;*
  (d) *All solutions of the associated homogeneous system tend to zero as $t \to +\infty$.*

Proof. As mentioned above, the stability of a linear system is given by the stability of the zero solution of the associated homogeneous system. Recall that the unique solution of a homogeneous linear system that satisfies the initial condition $x(t_0) = x_0$, is

$$x(t, t_0, x_0) = U(t)U(t_0)^{-1}x_0, \qquad (5.6)$$

where $U(t)$ is a fundamental matrix of the system. Then

$$\|x(t, t_0, x_0)\| \le \|U(t)\| \, \|U(t_0)^{-1}\| \, \|x_0\|, \qquad (5.7)$$

where by the norm $\|A\|$ of a square matrix $A = [a_{ij}]_{i,j=1,2,\ldots,n}$ we mean

$$\|A\| = \left(\sum_{i,j=1}^{n} a_{ij}^2 \right)^{\frac{1}{2}}.$$

(I) (b) ⇒ (a): If the fundamental matrix $U(t)$ is bounded on $\mathbb{R}_+$, then there is $M > 0$ such that $\|U(t)\| \leq M$ for all $t \in \mathbb{R}_+$. Consequently, if $\|x_0\| \leq \delta$, where $\delta = \delta(\varepsilon, t_0) := (M\|U(t_0)^{-1}\|)^{-1}\varepsilon$, then from (5.7),

$$\|x(t, t_0, x_0)\| \leq \varepsilon$$

for every $t \in [t_0, +\infty)$. This proves that the zero solution of the homogeneous system is stable, so the linear system is stable.

(b) ⇔ (c): Recall that any two fundamental matrices $U(t)$ and $V(t)$ of the same system are connected by the relation (2.23). From this relation we immediately see that one of the matrices is bounded on $\mathbb{R}_+$ if and only if the other one is.

(b) ⇒ (d): This implication immediately follows based on formula (5.6).

(d) ⇒ (b): The implication is trivial, since the columns of any fundamental matrix are solutions of the homogeneous system.

(a) ⇒ (b): From a chosen $\varepsilon = 1$, there exists a $\delta > 0$ such that

$$\text{if } \|x_0\| \leq \delta, \quad \text{then } \|x(t, 0, x_0)\| \leq 1 \quad \text{for all } t \geq 0.$$

We obtain n linearly independent solutions of the homogeneous system, bounded on $\mathbb{R}_+$, if we let x_0 be successively

$$\begin{bmatrix} \delta \\ 0 \\ \cdots \\ 0 \end{bmatrix}, \begin{bmatrix} 0 \\ \delta \\ \cdots \\ 0 \end{bmatrix}, \ldots, \begin{bmatrix} 0 \\ 0 \\ \cdots \\ \delta \end{bmatrix}.$$

Thus the matrix $U(t)$ having as columns these solutions is bounded on $\mathbb{R}_+$.

(II) The proof is based once again on the relation (5.6). □

5.3 Stability of Linear Systems with Constant Coefficients

In Section 2.10 it was shown that the stability of a linear differential system with constant coefficients depends on the eigenvalues of the matrix of coefficients. The result established there only for planar systems can be extended to the general case as shown in the next theorem.

Theorem 5.6 (Stability of constant coefficient linear systems). (a) *A linear system with constant coefficients is asymptotically stable if and only if all the eigenvalues of the matrix of coefficients have strictly negative real parts.*
(b) *A linear system with constant coefficients is stable if all the eigenvalues of the matrix of coefficients have nonpositive real parts and all the eigenvalues having the real part zero are simple.*
(c) *If at least one eigenvalue of the matrix of coefficients has the real part strictly positive, then the linear system is unstable.*

Proof. The results are direct consequences of Theorem 5.5 and Remark 2.16. □

A matrix for which all the eigenvalues have strictly negative real parts is said to be *Hurwitzian*, or a *Hurwitz*[2] *matrix*. According to the results from Section 2.10, a square matrix A of order 2 is Hurwitzian if and only if $\operatorname{tr} A < 0$ and $\det A > 0$.

Example 5.7. Consider the systems

$$\text{(i)} \begin{cases} x_1' = 0 \\ x_2' = 0 \end{cases} \qquad \text{(ii)} \begin{cases} x_1' = 0 \\ x_2' = x_1 \end{cases}.$$

The general solution for (i) is $[C_1, C_2]^T$, and for (ii) is $[C_1, C_1 t + C_2]^T$. Hence the first system is stable (but not asymptotically stable), while the second system is unstable. Notice that both systems have the same characteristic equation $r^2 = 0$ with double root $r = 0$. The first system is an example that shows that the condition from Theorem 5.6 (b) is only sufficient for stability, but not necessary.

Example 5.8. Check for stability the following systems:

$$\text{(a)} \begin{cases} x' = y + t \\ y' = z - 1 \\ z' = -2y - 2z\,; \end{cases} \qquad \text{(b)} \begin{cases} x' = x + y \\ y' = y + az \\ z' = z + x\,, \end{cases}$$

where a is a real parameter.

Solution. (a) The characteristic equation $\det(A - rI) = 0$ is $r^3 + 2r^2 + 2r = 0$ and its roots are $r_1 = -1 - i$, $r_2 = -1 + i$ and $r_3 = 0$. One has $\operatorname{Re} r_1 = \operatorname{Re} r_2 = -1 < 0$, $\operatorname{Re} r_3 = 0$, and the root r_3 is simple. Hence the system is stable, but not asymptotically stable.

(b) The characteristic equation is $(1 - r)^3 + a = 0$. Hence the eigenvalues are $r_1 = r_2 = r_3 = 1 + \sqrt[3]{a}$. Thus the system is asymptotically stable for $a < -1$, and unstable for $a > -1$.

5.4 Stability of Solutions of Nonlinear Systems

In what follows we discuss the stability of the zero solution for the nonlinear differential system

$$x' = Ax + p(t, x)\,, \tag{5.8}$$

where A is a square matrix of order n, and $p : \mathbb{R}_+ \times B_a \to \mathbb{R}^n$ is continuous in (t, x), locally Lipschitz continuous in x, and satisfies $p(t, 0) = 0$ for all $t \in \mathbb{R}_+$. Here by B_a we have denoted the open ball of $\mathbb{R}^n$ of radius a centered at the origin. Such a system is called a *perturbed linear system*, where p is the *perturbation term*. We expect that the stability property of the linear system $x' = Ax$ is passed on the zero solution of the

---

[2] Adolf Hurwitz (1859–1919)

perturbed system if the perturbation term p is sufficiently small. To show this we need the following lemma:

Lemma 5.9. *If the matrix A is Hurwitzian, then there exist two positive constants M and ω such that*

$$\left\| e^{tA} \right\| \leq M e^{-\omega t} \tag{5.9}$$

for every $t \geq 0$.

Proof. If A is Hurwitzian, then the real parts of its eigenvalues are strictly negative. Hence we may choose a number $\omega > 0$ such that

$$\operatorname{Re} r < -\omega$$

for every eigenvalue r of A. Note that the eigenvalues r of A and the eigenvalues $\hat{r}$ of the matrix $A + \omega I$ are in the relation $\hat{r} = r + \omega$. Hence

$$\operatorname{Re} \hat{r} = \operatorname{Re} r + \omega < 0,$$

and so the matrix $A + \omega I$ is also Hurwitzian. Then the linear system $x' = (A + \omega I)x$ is asymptotically stable and from Theorem 5.5, the fundamental matrix $e^{t(A+\omega I)}$ tends to the zero matrix as $t \to +\infty$. Consequently, there is a constant $M > 0$ such that

$$\left\| e^{t(A+\omega I)} \right\| \leq M \tag{5.10}$$

for every $t \geq 0$. Conversely,

$$e^{t(A+\omega I)} = e^{tA} e^{\omega t I} = e^{tA} \left(e^{\omega t} I \right) = e^{\omega t} e^{tA}.$$

Thus

$$\left\| e^{t(A+\omega I)} \right\| = e^{\omega t} \left\| e^{tA} \right\|. \tag{5.11}$$

Now (5.10) and (5.11) yield (5.9). □

Theorem 5.10. *Let the matrix A be Hurwitzian and the numbers M and ω be as in (5.9). If there exists a positive number $L < \omega/M$ such that*

$$\|p(t, x)\| \leq L \|x\|$$

for every $(t, x) \in \mathbb{R}_+ \times B_a$, then the zero solution of the perturbed system (5.8) is asymptotically stable.

Proof. Let $(t_0, x_0) \in \mathbb{R}_+ \times B_a$ and $x(t, t_0, x_0)$ be the unique saturated solution of the system (5.8) that satisfies the initial condition $x(t_0) = x_0$. Let $[t_0, t_+)$ be the domain of $x(t, t_0, x_0)$ to the right of t_0. From the representation formula (2.28), one has

$$x(t, t_0, x_0) = e^{(t-t_0)A} x_0 + \int_{t_0}^{t} e^{(t-s)A} p(s, x(s, t_0, x_0)) \, ds, \quad t_0 \leq t < t_+.$$

Then, by virtue of Lemma 5.9, it follows that

$$\|x(t, t_0, x_0)\| \le Me^{-\omega(t-t_0)} \|x_0\| + LM \int_{t_0}^{t} e^{-\omega(t-s)} \|x(s, t_0, x_0)\| \, ds \,.$$

Multiplying both sides of this inequality by $e^{\omega t}$ and denoting $\phi(t) = e^{\omega t}\|x(t, t_0, x_0)\|$, we obtain

$$\phi(t) \le Me^{\omega t_0} \|x_0\| + LM \int_{t_0}^{t} \phi(s) \, ds \,.$$

From Gronwall's inequality we deduce

$$\phi(t) \le Me^{\omega t_0 + LM(t-t_0)} \|x_0\| \,.$$

Then

$$\|x(t, t_0, x_0)\| \le Me^{(LM-\omega)(t-t_0)} \|x_0\|, \quad t_0 \le t < t_+, \tag{5.12}$$

where $LM - \omega < 0$. From (5.12) it follows that if $\|x_0\| \le a/(2M)$, then

$$\|x(t, t_0, x_0)\| \le \frac{a}{2}, \quad t_0 \le t < t_+,$$

which shows that the graph of the solution $x(t, t_0, x_0)$ does not approach the boundary of $\mathbb{R}_+ \times B_a$ as $t \to t_+$. Since the solution is also bounded on $[t_0, t_+)$ taking values in the ball B_a, it cannot blow up in finite time. Thus, according to Theorem 4.32, the solution $x(t, t_0, x_0)$ is defined on the whole semiline $[t_0, +\infty)$, i.e. $t_+ = +\infty$. Then the inequality (5.12) holds for every $t \in [t_0, +\infty)$, and since $e^{(LM-\omega)(t-t_0)} \to 0$ as $t \to +\infty$, it immediately implies that the zero solution is asymptotically stable. □

Corollary 5.11. *If the matrix A is Hurwitzian and there exists a function* $q \colon \mathbb{R}_+ \to \mathbb{R}_+$ *such that*

$$\|p(t, x)\| \le q(\|x\|) \tag{5.13}$$

for every $(t, x) \in \mathbb{R}_+ \times B_a$, *and*

$$\lim_{\tau \to 0} \frac{q(\tau)}{\tau} = 0, \tag{5.14}$$

then the zero solution of the system (5.8) is asymptotically stable.

Proof. Let M and ω be two positive numbers such that (5.9) holds for every $t \ge 0$. Fix $L > 0$ with the property $L < \omega/M$. From (5.14) it follows that there exists $a_L \in (0, a]$ such that

$$q(\tau) \le L\tau \text{ for every } \tau \in [0, a_L] \,.$$

Then from (5.13),

$$\|p(t, x)\| \le L \|x\|$$

for every $(t, x) \in \mathbb{R}_+ \times B_{a_L}$. Thus we are under the hypotheses of Theorem 5.10, with B_{a_L} in place of B_a. □

Corollary 5.11 allows us to establish a very useful asymptotic stability criterion for the zero solution of autonomous nonlinear systems of the form

$$x' = f(x). \tag{5.15}$$

Here it is assumed that $f: B_a \to \mathbb{R}^n$, $f = (f_1, f_2, \ldots, f_n)$, is a function of class C^1, with $f(0) = 0$. Let $f_x(0)$ be the Jacobian[3] matrix of f at 0, that is

$$f_x(0) := \left[\frac{\partial f_i(0)}{\partial x_j} \right]_{i,j=1,2,\ldots,n}.$$

Theorem 5.12 (Asymptotic stability criterion). *If the Jacobian matrix $A = f_x(0)$ is Hurwitzian, then the zero solution of the system (5.15) is asymptotically stable.*

Proof. Since f is of class C^1, one has

$$f(x) = f(0) + f_x(0)x + p(x) = f_x(0)x + p(x)$$

for every $x \in B_a$, where

$$\lim_{x \to 0} \frac{\|p(x)\|}{\|x\|} = 0.$$

Thus we are in the conditions of Corollary 5.11 with

$$q(\tau) = \sup \{\|p(x)\| : x \in B_a, \|x\| = \tau\}$$

for $0 \le \tau \le a$. □

Remark 5.13. For the study of stability of an arbitrary equilibrium (stationary solution) x^* of the autonomous system (5.15) one makes the change of variable $y := x - x^*$ and turns the system in x into a system in y, namely $y' = g(y)$, where $g(y) := f(y + x^*)$. One has

$$g_y(0) = f_x(x^*),$$

and the conclusion follows: if the Jacobian matrix $f_x(x^*)$ is Hurwitzian, then the equilibrium x^* is asymptotically stable.

This method of stability analysis, based on the linear approximation of the function f in the vicinity of an equilibrium, is called *Lyapunov's linearization method*.

Remark 5.14. In the case of a single autonomous differential equation, i.e. for $n = 1$, the Jacobian matrix reduces to the derivative $f'(x^*)$ and thus the property of being Hurwitzian reads as $f'(x^*) < 0$.

Therefore, if $f(x^*) = 0$ and $f'(x^*) < 0$, then the equilibrium $x = x^*$ of the equation $x' = f(x)$ is asymptotically stable. It can be proved that if by the contrary $f'(x^*) > 0$, then the equilibrium x^* is unstable. In the case where $f'(x^*) = 0$, using this method, one cannot decide the stability of x^*.

---

[3] Carl Gustav Jacob Jacobi (1804–1851)

5.4 Stability of Solutions of Nonlinear Systems

Example 5.15. Check for stability the equilibria of the following equations:

(a) $x' = x(x-1)$; (b) $x' = -x^2$; (c) $x' = -\frac{1}{2}x^3$.

Solution. (a) The equilibria are $x_1 = 0$ and $x_2 = 1$. If $f(x) = x(x-1)$, then $f'(0) = -1 < 0$ and $f'(1) = 1 > 0$. Hence the equilibrium $x_1 = 0$ is asymptotically stable, while $x_2 = 1$ is unstable.

(b) There is only one equilibrium, $x = 0$. If $f(x) = -x^2$, then $f'(0) = 0$, so using the linearization method we cannot decide the stability of this equilibrium. We try to do it directly using the definitions of stability. Solving the equation with the initial condition $x(0) = x_0$, we obtain

$$x(t, x_0) = \frac{x_0}{1 + tx_0}.$$

We see that no matter how close x_0 is to zero, if $x_0 < 0$, then $x(\cdot, x_0)$ only exists on the finite interval $[0, -1/x_0)$. Hence the solution $x = 0$ is not stable.

(c) In this case, again $x = 0$ is the unique equilibrium of the differential equation, and $f'(0) = 0$, where $f(x) = -x^3/2$. Solving as above gives

$$x(t, x_0) = \frac{x_0}{\sqrt{1 + tx_0^2}}.$$

This solution is defined on $\mathbb{R}_+$ and tends to zero as $t \to +\infty$. It follows that the equilibrium $x = 0$ is asymptotically stable.

Example 5.16. The equilibria of the logistic equation

$$x' = rx\left(1 - \frac{x}{k}\right)$$

are 0 and k. If $f(x) = rx(1-x/k)$, then $f'(x) = r - 2rx/k$, $f'(0) = r > 0$, and $f'(k) = -r < 0$. Thus the zero solution is unstable and the solution $x = k$ is asymptotically stable. In an ecological context, this means that small perturbations to the equilibrium abundance of a population do not essentially modify the evolution of the species, which in time returns to its original equilibrium state.

Example 5.17. Determine the type of stability of the stationary solutions of the differential system

$$\begin{cases} x' = -x^2 + y \\ y' = -y^2 + x. \end{cases}$$

Solution. The stationary solutions are obtained by solving the algebraic system $-x^2 + y = 0$, $-y^2 + x = 0$. They are $(0, 0)$ and $(1, 1)$. The Jacobian matrix is

$$\begin{bmatrix} -2x & 1 \\ 1 & -2y \end{bmatrix}.$$

For $x = y = 0$ it is not Hurwitzian, but for $x = y = 1$ it is so. Thus, the equilibrium $(1, 1)$ is asymptotically stable.

5.5 Method of Lyapunov Functions

In the previous section we presented Lyapunov's linearization method (also called the *indirect method of Lyapunov*) for the stability of equilibria of the autonomous differential system

$$x' = f(x), \tag{5.16}$$

where $f: D_0 \to \mathbb{R}^n$, $f = (f_1, f_2, \ldots, f_n)$ is a function of class C^1 on the open neighborhood D_0 of 0, and $f(0) = 0$.

Recall that for autonomous systems, any stable equilibrium is necessarily uniformly stable and any asymptotically stable equilibrium is uniformly asymptotically stable. Thus the equilibrium $x = 0$ is stable if

$$\text{for each } \varepsilon > 0, \text{ there exists } \delta = \delta(\varepsilon) > 0 \text{ such that} \tag{5.17}$$

$$\|x_0\| \le \delta \quad \text{implies} \quad \|x(t, x_0)\| \le \varepsilon \text{ for all } t \ge 0,$$

and is asymptotically stable if it is stable. In addition

$$\text{there exists } \eta > 0 \text{ such that}$$

$$\|x_0\| \le \eta \quad \text{implies} \quad \lim_{t \to +\infty} x(t, x_0) = 0.$$

The basic idea of the method of Lyapunov functions, also called *Lyapunov's direct method*, is inspired by the behavior of physical systems that lose energy along their evolution towards a final resting state. However, there are systems for which the concept of energy is not applicable. Then, more generally, instead of the energy, one may consider a continuous function $V: B_a \to \mathbb{R}$, on an open ball $B_a = \{x: \|x\| < a\} \subseteq D_0$, such that $V(0) = 0$, $V(x) > 0$ for $x \ne 0$ and $V(x(., x_0))$ is nondecreasing on $[0, t_+)$ for every saturated solution $x(., x_0)$ of the system (5.16) with $x_0 \in B_a$. If such a function V exists, then the equilibrium 0 is (uniformly) stable. To prove this assertion, let $\varepsilon > 0$ be given, and assume without loss of generality that $\overline{B}_\varepsilon \subset B_a$. Let $m = \min\{V(x): \|x\| = \varepsilon\}$. Clearly, $m > 0$. Using the continuity of V at 0, we may find $\delta \in (0, \varepsilon)$ such that $V(x) < m$ for $\|x\| \le \delta$. For $\|x_0\| \le \delta$, from

$$V(x(t, x_0)) \le V(x(0, x_0)) = V(x_0) < m, \quad 0 \le t < t_+$$

we see that the trajectory $x(t, x_0)$ ($t \in [0, t_+)$ does not leave the open ball B_ε. Since $\overline{B}_\varepsilon \subset D_0$, we have $t_+ = +\infty$ and so $\|x(t, x_0)\| < \varepsilon$ for all $t \in \mathbb{R}_+$. Therefore, the equilibrium $x = 0$ is stable as claimed.

A sufficient condition for $V(x(., x_0))$ to be nondecreasing on $[0, t_+)$ for every saturated solution $x(., x_0)$ of the system (5.16) with $x_0 \in B_a$, is that $V \in C^1(B_a)$ and

$$\nabla V(x) \cdot f(x) \le 0 \quad \text{for all } x \in B_a,$$

where $\nabla V(x)$ is the vector

$$\left(\frac{\partial V(x)}{\partial x_1}, \frac{\partial V(x)}{\partial x_2}, \ldots, \frac{\partial V(x)}{\partial x_n} \right).$$

Indeed, one has

$$\frac{d}{dt}V(x(t,x_0)) = \nabla V(x(t,x_0)) \cdot x'(t,x_0)$$
$$= \nabla V(x(t,x_0)) \cdot f(x(t,x_0)) \leq 0,$$

which shows that the function $V(x(\cdot,x_0))$ is nondecreasing on $[0,t_+)$.

Definition 5.18. A function $V: G \to \mathbb{R}$, where $G \subseteq D_0$ is open and $0 \in G$, is said to be a *Lyapunov function* on G for the system (5.16) if the following conditions hold:
(i) $V \in C^1(G)$, $V(0) = 0$;
(ii) $V(x) > 0$ for every $x \in G \setminus \{0\}$;
(iii) $\nabla V(x) \cdot f(x) \leq 0$ for every $x \in G$.

Therefore one has the following stability criterion:

Theorem 5.19. (a) *If the system* (5.16) *has a Lyapunov function on a neighborhood of the origin, then the equilibrium $x = 0$ is stable.*
(b) *If the system* (5.16) *has a Lyapunov function on a neighborhood of the origin such that*

$$\nabla V(x) \cdot f(x) < 0 \qquad (5.18)$$

for $x \neq 0$, then the equilibrium $x = 0$ is asymptotically stable.

Proof. Obviously if V is a Lyapunov function on a neighborhood G of the origin, then its restriction to any open ball $B_a \subseteq G$ is a Lyapunov function on B_a. Let us fix such a ball B_a. The assertion (a) has been already proved. To prove (b) we only need to show that there exists $\eta \in (0, a)$ such that

$$\lim_{t \to +\infty} x(t, x_0) = 0 \qquad (5.19)$$

for $\|x_0\| \leq \eta$. Let us fix any number $\varepsilon_0 \in (0, a)$, let $\delta_0 \in (0, a)$ be the corresponding number given by (5.17), and let $\eta = \min\{\varepsilon_0, \delta_0\}$. Then for $\|x\| \leq \eta$,

$$V(x) \geq \omega(\|x\|) \qquad (5.20)$$

where

$$\omega(\tau) = \min\{V(x): \tau \leq \|x\| \leq \eta\}.$$

Obviously, ω is continuous, nondecreasing, $\omega(0) = 0$ and $\omega(\tau) > 0$ for $\tau \in (0, \eta]$. In view of (5.20), for (5.19) to hold, it suffices that

$$\lim_{t \to +\infty} V(x(t,x_0)) = 0. \qquad (5.21)$$

Assume the contrary. Then $\lim_{t \to +\infty} V(x(t,x_0)) = \kappa > 0$. Hence $V(x(t,x_0)) \geq \kappa$ for all $t \in \mathbb{R}_+$. Using the continuity of V at 0, we may find $\delta \in (0, \eta)$ such that $V(x) < \kappa$ for $\|x\| < \delta$. Hence $\|x(t,x_0)\| \geq \delta$ for all $t \in \mathbb{R}_+$. Let $\gamma = \min\{-\nabla V(x) \cdot f(x): \delta \leq \|x\| \leq \eta\}$. Then $\gamma > 0$ and

$$-\frac{d}{dt}V(x(t,x_0)) = -\nabla V(x(t,x_0)) \cdot f(x(t,x_0)) \geq \gamma.$$

Integrating from 0 to t yields
$$-V(x(t, x_0)) + V(x_0) \geq \gamma t,$$
or equivalently
$$0 < V(x(t, x_0)) \leq V(x_0) - \gamma t.$$
Letting $t \to +\infty$ yields the contradiction $0 \leq -\infty$. Therefore (5.21) holds. □

Notice that there are cases for which the linearization method does not apply, but the method of Lyapunov functions does. Such a case is given by the next example.

Example 5.20. Show that the system
$$\begin{cases} x_1' = -x_1^3 + x_2 \\ x_2' = -x_1 - x_2^3 \end{cases}$$
admits a Lyapunov function of the form $c_1 x_1^2 + c_2 x_2^2$, and determine whether the zero solution is stable or asymptotically stable.

Solution. Let $V(x_1, x_2) = c_1 x_1^2 + c_2 x_2^2$. Then $\nabla V = (2c_1 x_1, 2c_2 x_2)$ and
$$\nabla V(x) \cdot f(x) = 2c_1 x_1 (-x_1^3 + x_2) + 2c_2 x_2 (-x_1 - x_2^3)$$
$$= -2c_1 x_1^4 - 2c_2 x_2^4 + 2x_1 x_2 (c_1 - c_2).$$
Hence, if we let $c_1 = c_2 =: c > 0$, we obtain
$$\nabla V(x) \cdot f(x) = -2c(x_1^4 + x_2^4) < 0$$
for $x = (x_1, x_2) \neq 0$. Therefore, any function of the form $c(x_1^2 + x_2^2)$ with $c > 0$ is a Lyapunov function of the system that satisfies (5.18), and consequently the zero solution is asymptotically stable. However, the Jacobian matrix at the origin is
$$\begin{bmatrix} 0 & 1 \\ -1 & 0 \end{bmatrix}$$
and has the eigenvalues $r = \pm i$. Hence it is not Hurwitzian and Theorem 5.12 is not applicable.

Example 5.21. Consider the linear system
$$\begin{cases} x' = -2x + y \\ y' = -x - 3y. \end{cases}$$
Use direct and indirect Lyapunov methods to prove that the system is asymptotically stable.

Solution. Direct method: The function $V(x, y) = x^2 + y^2$ is a Lyapunov function of the system with
$$\nabla V(x, y) \cdot f(x, y) = -4x^2 - 6y^2 < 0$$
for $(x, y) \neq (0, 0)$. Hence the system is asymptotically stable.

Indirect method: The characteristic equation is $r^2 + 5r + 7 = 0$ with roots $r_1, r_2 = (-5 \pm i\sqrt{3})/2$. Since $\operatorname{Re} r_1 = \operatorname{Re} r_2 < 0$, the system is asymptotically stable.

Example 5.22. Consider the *gradient system* $x' = -\nabla F(x)$, where $F \colon B_a \subset \mathbb{R}^n \to \mathbb{R}$ is of class C^2. Assume that $F(0) = 0$, and that $F(x) > 0$ and $\nabla F(x) \neq 0$ for all $x \in B_a \setminus \{0\}$, i.e. $x = 0$ is an isolated minimum point of F. Show that the zero solution is asymptotically stable.

Solution. Denote $f(x) = -\nabla F(x)$ and $V(x) = F(x)$. One has
$$\nabla V(x) \cdot f(x) = -\|\nabla F(x)\|^2 < 0$$
for all $x \in B_a \setminus \{0\}$. Hence $F(x)$ is a Lyapunov function of the system that satisfies (5.18). Therefore, the solution $x = 0$ is asymptotically stable.

Example 5.23. Show that the zero solution of the *Hamiltonian system*
$$\begin{cases} x' = \frac{\partial H}{\partial y}(x, y) \\ y' = -\frac{\partial H}{\partial x}(x, y) \end{cases}$$
is stable. Here the function $H \colon \mathbb{R}^2 \to \mathbb{R}$ is continuously differentiable, with $H(0, 0) = 0$, $H(x, y) > 0$ for $(x, y) \neq (0, 0)$, and $\nabla H(0, 0) = (0, 0)$.

Solution. The function H is a Lyapunov function of the system.

Assume now that we are interested into the stability property of an arbitrary equilibrium $x = x^*$ of the system (5.16). Hence
$$x^* \in D_0 \quad \text{and} \quad f(x^*) = 0.$$
The substitution $y = x - x^*$ leads to the system $y' = g(y)$, where $g(y) = f(y + x^*)$ for $y \in D_0 - x^*$, and the stability of $x = x^*$ is equivalent to that of the zero solution $y = 0$ of the new system. Now observe that the new system admits a Lyapunov function $V(y)$ on a neighborhood G_0 of 0, if and only if there exists a function $W(x)$ on the neighborhood $G_{x^*} := x^* + G_0$ of x^*, such that:
(i) $W \in C^1(G_{x^*})$, $W(x^*) = 0$;
(ii) $W(x) > 0$ for every $x \in G_{x^*} \setminus \{x^*\}$;
(iii) $\nabla W(x) \cdot f(x) \leq 0$ for every $x \in G_{x^*}$.

Between the two functions V and W one has the relation
$$V(y) = W(x^* + y).$$
A function $W(x)$ satisfying the above properties is also called a *Lyapunov function* and its existence on a neighborhood of x^* is sufficient for the equilibrium $x = x^*$ to be stable. The additional condition
$$\nabla W(x) \cdot f(x) < 0 \quad \text{for every } x \in G_{x^*} \setminus \{x^*\}$$
is sufficient for $x = x^*$ to be asymptotically stable.

Example 5.24. The function

$$W(x, y) = y - \frac{a}{b} - \frac{a}{b}\ln\left(\frac{by}{a}\right) + \frac{c}{b}\left[x - \frac{d}{c} - \frac{d}{c}\ln\left(\frac{cx}{d}\right)\right]$$

is a Lyapunov function with respect to the equilibrium solution $(d/c, a/b)$ of the Lotka–Volterra predator-prey system

$$\begin{cases} x' = ax - bxy \\ y' = cxy - dy, \end{cases}$$

where $a, b, c, d > 0$. Hence the stationary solution $(d/c, a/b)$ is stable. Note that the conclusion cannot be derived using the linearization method.

Theorem 5.19 only gives sufficient conditions of stability, and there are significant cases where it fails in spite of evident stability properties shown by the phase portrait or just by common sense. Such a case is given by the damped pendulum equation

$$x'' + a\sin x + bx' = 0,$$

where $a, b > 0$, which is equivalent to the differential system

$$\begin{cases} x' = y \\ y' = -a\sin x - by. \end{cases} \tag{5.22}$$

It can be seen that the total (kinetic + potential) energy associated with the pendulum without friction,

$$\frac{1}{2}x'^2 + \int_0^x a\sin x\, dx = \frac{1}{2}x'^2 + a(1 - \cos x),$$

gives us the Lyapunov function of (5.22) on $(-\pi, \pi) \times \mathbb{R}$, namely

$$V(x, y) = a(1 - \cos x) + \frac{1}{2}y^2,$$

for which

$$\nabla V(x, y) \cdot f(x, y) = (a\sin x,\ y) \cdot (y,\ -a\sin x - by) = -by^2.$$

Hence from Theorem 5.19 (a), the zero equilibrium is stable, but since $\nabla V(x, y) \cdot f(x, y) \not< 0$ for all $(x, y) \neq (0, 0)$, we cannot apply Theorem 5.19 (b) and conclude the evident asymptotic stability of the origin. A clarification is given by the next result.

Theorem 5.25. *If the system (5.16) has a Lyapunov function V on a neighborhood G of the origin, and the set*

$$S := \{x \in G : \nabla V(x) \cdot f(x) = 0\}$$

contains no trajectory of the system except the trivial trajectory $x(t) = 0$ for $t \geq 0$, then the equilibrium $x = 0$ is asymptotically stable.

Proof. Let $B_a \subset G$. According to Theorem 5.19 (a), the zero equilibrium is stable. Hence if we fix any number $\varepsilon_0 \in (0, a)$, then there exists $\delta_0 \in (0, a)$ such that (5.17) holds. Let $\eta = \min\{\varepsilon_0, \delta_0\}$. Then for every x_0 with $\|x_0\| \leq \eta$, one has $\|x(t, x_0)\| \leq \eta$ for all $t \geq 0$. We claim that in addition $x(t, x_0) \to 0$ as $t \to +\infty$. To prove this, first note that since $V(x(t, x_0))$ is decreasing, there is a number c such that $c = \lim_{t \to +\infty} V(x(t, x_0))$. Let y be any limit point of the trajectory $x(., x_0)$. Hence there exists a sequence $t_k \to +\infty$ such that $x(t_k, x_0) \to y$ as $k \to \infty$. Then $V(y) = c$. On the other hand, each point $x(t, y)$ of the trajectory $x(., y)$, for $t > 0$, is also a limit point of the trajectory $x(., x_0)$ since

$$x(t + t_k, x_0) = x(t, x(t_k, x_0)) \to x(t, y)$$

as $k \to \infty$. Hence, as above, $V(x(t, y)) = c$ for every $t \geq 0$. Consequently,

$$0 = \frac{d}{dt} V(x(t, y)) = \nabla V(x(t, y)) \cdot f(x(t, y)), \quad t \geq 0.$$

This shows that the entire trajectory $x(t, y)$ for $t \geq 0$ is contained in S. Hence, by hypothesis, $x(t, y) = 0$ for all $t \geq 0$. In particular one has $y = 0$. Therefore $y = 0$ is the unique limit point of the trajectory $x(., x_0)$, that is $\lim_{t \to +\infty} x(t, x_0) = 0$, as claimed. □

Remark 5.26. If the Lyapunov function V satisfies (5.18), then the condition that S contains no trajectory of the system except the trivial trajectory $x(t) = 0$ for $t \geq 0$ is trivially satisfied. Hence Theorem 5.19 (b) is a particular case of Theorem 5.25.

Example 5.27. Coming back to the damped pendulum equation considered above, let us see that $S = \{(x, 0): -\pi < x < \pi\}$ and that if a trajectory $(x(t), y(t))$ is entirely contained in S, then $y(t) = 0$ for all $t \geq 0$, and from the first equation in (5.22), $x(t)$ is constant. Next using the second equation we deduce that $\sin x(t) = 0$. Hence $x(t) = 0$ for all $t \geq 0$. Therefore we are under the assumptions of Theorem 5.25. Hence the zero equilibrium is asymptotically stable, as physically inferred.

5.6 Globally Asymptotically Stable Systems

According to the definition, if the equilibrium point $x = 0$ of the system (5.16) is asymptotically stable, then (5.19) holds for all x_0 in some neighborhood of zero. The set of all initial data $x_0 \in D_0$ for which (5.19) holds is called the *basin of attraction* of 0. If D_0 is the entire space $\mathbb{R}^n$ and $x = 0$ is the unique equilibrium point of the system, one may be interested in the case when the basin of attraction of $x = 0$ is the entire space. In such a case, we say that the equilibrium $x = 0$ is *globally asymptotically stable*, or that the system is globally asymptotically stable. The next theorem is a criterion of global asymptotic stability.

Theorem 5.28. *Assume that $D_0 = \mathbb{R}^n$ and $f(0) = 0$. If the system (5.16) has a Lyapunov function V on $\mathbb{R}^n$ such that the set $S := \{x \in \mathbb{R}^n : \nabla V(x) \cdot f(x) = 0\}$ contains no trajectory* of the system except the trivial trajectory $x(t) = 0$ for $t \geq 0$, and V is radially unbounded, i.e.

$$V(x) \to +\infty \quad \text{as} \quad \|x\| \to +\infty,$$

then the system is globally asymptotically stable.

Proof. The radial unboundedness of V guarantees that any solution $x(t, x_0)$ is bounded on its maximum interval of definition $[0, t_+)$. Indeed, otherwise there would exists a sequence $t_k \to t_+$ with $\|x(t_k, x_0)\| \to +\infty$, hence $V(x(t_k, x_0)) \to +\infty$, which is impossible since $V(x(t_k, x_0)) \leq V(x(0, x_0)) = V(x_0)$. Next, according to the theorem on the global existence and uniqueness of solutions for the Cauchy problem, being bounded functions, all trajectories have to be defined on the entire semiline $\mathbb{R}_+$. Finally the conclusion that $x(t, x_0) \to 0$ as $t \to +\infty$ is obtained as in the proof of Theorem 5.25. □

Example 5.29. Consider the equation $x'' + ax' + g(x) = 0$, where $a > 0$, $g: \mathbb{R} \to \mathbb{R}$ is continuous, $xg(x) > 0$ for $x \neq 0$ and $G(x) = \int_0^x g(s)\, ds \to +\infty$ as $|x| \to +\infty$. The equation is a model for a spring-mass motion with friction. The equivalent system is

$$\begin{cases} x' = y \\ y' = -g(x) - ay. \end{cases}$$

The total energy $V(x, y) = y^2/2 + G(x)$ is a radially unbounded Lyapunov function on $\mathbb{R}^2$, with $\nabla V(x, y) \cdot f(x, y) = -ay^2 \leq 0$. In this case $S = \{(x, 0): x \in \mathbb{R}\}$. If a trajectory $(x(t), y(t))$ is entirely contained in S for $t \geq 0$, then $y(t) = 0$ and from the first equation of the system $x(t) = c$, a constant. Next from the second equation we must have $c = 0$. Hence the only trajectory contained in S is the origin. Therefore, the system is globally asymptotically stable.

Finally we note that the above direct stability analysis can be extended to nonautonomous systems.

Chapter 6
Differential Systems with Control Parameters

Frequently, differential equations and systems contain one or more parameters that often are assumed to be constant, but which can be dependent themselves on the independent variable, or even on the state variable. For example, the equation

$$x' = ax\left(1 - \frac{x}{k}\right) - \lambda x \tag{6.1}$$

modeling the dynamics of a species in the presence of harvesting, contains three parameters: a, the growth rate per capita; k, the carrying capacity; and λ, the effort of harvesting. If harvesting is constant in time, then the parameter λ is a constant, while if the effort of harvesting changes, for instance seasonally, then λ is a function of time.

A natural question to ask is how the qualitative properties of differential systems change as one or several parameters of the system are varied, and how can we find those values of the parameters that make the system behave conveniently? The parameters that are varied are said to be *control parameters* to distinguish them from those parameters that remain fixed. In fact, when one or more parameters of a system are varied, we are in the presence of a family of differential systems parametrized by the control parameters.

6.1 Bifurcations

In this and the next section we shall consider differential equations and systems with only one control parameter. Roughly speaking, a *bifurcation* is a qualitative change in the system's behavior caused by the change of the parameter values. A value of the control parameter at which the system's behavior changes is said to be a *bifurcation point*. Most commonly, bifurcations are concerned with the number of equilibria and their stability properties, and in a strict sense, the term bifurcation is used only to denote points where the stability of the system changes.

Let us consider a few examples that exhibit various bifurcation situations.

Example 6.1. Consider the equation (6.1), where $a, k > 0$ are fixed constants and λ is a real parameter. The equation has two equilibria: $x_1 = 0$ and $x_2 = (a - \lambda)k/a$. If $f(x) = ax(1 - x/k) - \lambda x$, then $f'(x) = a - \lambda - 2ax/k$. Hence $f'(x_1) = a - \lambda$ and $f'(x_2) = -(a - \lambda)$. Consequently, if $\lambda < a$, the equilibrium 0 is unstable and the positive equilibrium x_2 is asymptotically stable; while if $\lambda > a$, the equilibrium 0 is asymptotically stable and the (negative) equilibrium x_2 is unstable. Therefore, the value $\lambda = a$ is a bifurcation point. From the ecologic point of view, if the harvesting effort is less than the growth rate, the species will not die out, but when the harvesting effort equals the growth rate there is a dramatic change leading to species extinction.

Example 6.2. The equation $x' = \lambda x - x^3$ has, for $\lambda \leq 0$, only the equilibrium $x = 0$, which is asymptotically stable; while for $\lambda > 0$, it has three equilibria: $x_1 = 0$ (unstable), and $x_2, x_3 = \pm\sqrt{\lambda}$ (asymptotically stable). Hence $\lambda = 0$ is a bifurcation point.

In the case of one-parameter differential equations, it is common to illustrate what is going on as the control parameter varies with the so-called *bifurcation diagram*. It contains the plot of the equilibria as functions of the control parameter. Commonly, the branches of the bifurcation diagram are labeled by (s) and (u) showing after case, stability or instability. Thus, for Example 6.2, the bifurcation diagram contains the graphs of the functions: $x_1(\lambda) = 0$, labeled (s) for $\lambda \leq 0$, and (u) for $\lambda > 0$; and $x_2(\lambda), x_3(\lambda) = \pm\sqrt{\lambda}$ for $\lambda > 0$, labeled (s). On account of the diagram's appearance, the bifurcation is said to be a *pitchfork bifurcation*.

Example 6.3. In Section 4.1, the system

$$\begin{cases} x' = \dfrac{ax}{1 + b(x + y)} - cx \\ y' = \dfrac{Ax}{1 + B(x + y)} - Cy \end{cases}$$

was considered as a model for the hematologic normal-leukemic cell dynamics. Here $x(t)$ and $y(t)$ represent the number of normal and leukemic cells respectively; a, A are the growth rates of the two cell populations; c, C are the cell death rates; and b, B stand for the sensibility of the two types of cells with respect to the crowding effect in the bone marrow microenvironment. The parameters a, b, c characterize the human species, while the parameters A, B, C vary from one person to another. It is natural to assume that for both types of cells, the cell death rate is less than the growth rate, i.e. $c < a$ and $C < A$. It is easy to see that the system has three equilibria: $(0, 0)$, $(d, 0)$ and $(0, D)$, where

$$d = \frac{1}{b}\left(\frac{a}{c} - 1\right), \quad D = \frac{1}{B}\left(\frac{A}{C} - 1\right).$$

Using the linearization method of stability, that is, making use of the Jacobian matrix associated with the system, we can arrive with a bit work at the following conclusions:
(a) the equilibrium $(0, 0)$ is always unstable;
(b) the equilibrium $(d, 0)$ is asymptotically stable and $(0, D)$ is unstable if $D < d$;
(c) the equilibrium $(0, D)$ is asymptotically stable and $(d, 0)$ is unstable if $D > d$.

From a medical point of view, the situation (b) corresponds to the healthy state of a person, while (c) shows acute leukemic disease. Indeed, if $(d, 0)$ is asymptotically stable, then the number $x(t)$ of normal cells tends to the maximum level of abundance d, and the leukemic cell population $y(t)$ tends to zero. Similarly, if $(0, D)$ is asymptotically stable, then the leukemic cells proliferate towards an upper level D to the detriment of the normal cells, which are ultimately eliminated. For this example we may consider D as a control parameter even though it is not directly involved in the system. Then the value $D = d$ is a bifurcation point of the system. Of course, instead of D we may take as control parameter any one of the parameters A, B, C. For instance, if A is taken as a control parameter, then the bifurcation point is

$$A = C\left[1 + \frac{B}{b}\left(\frac{a}{c} - 1\right)\right],$$

which is immediately obtained from the equality $D = d$.

Example 6.4. Consider the Lotka–Volterra system where both the prey and the predator are harvested proportional to their population sizes

$$\begin{cases} x' = ax - bxy - \lambda x \\ y' = cxy - dy - \lambda y \,. \end{cases}$$

The equilibria are $(0, 0)$ and (x^*, y^*), where $x^* = (d + \lambda)/c$ and $y^* = (a - \lambda)/b$. Using the linearization method one can find that in the case $\lambda > a$, the equilibrium $(0, 0)$ is asymptotically stable, meaning the extinction of the two species. However, if $\lambda < a$, then $(0, 0)$ is unstable and the positive stationary solution (x^*, y^*) is stable as shown in Example 5.24 using Lyapunov's direct method (in the expression of Lyapunov's function, take this time $a - \lambda$, $d + \lambda$ instead of a and d). It is worth underlining here that the bifurcation point coincides with the growth rate of the prey population.

6.2 Hopf Bifurcations

In the examples from the previous section, the stability of an equilibrium point depending on a parameter is lost at bifurcation and recovered by one or more equilibria. In the case of Hopf[1] bifurcation, the transfer of stability occurs between equilibria and periodic solutions.

To be more explicit, let us consider in this section only planar autonomous differential systems and adopt some specific terminology on dynamical systems. Let

$$x' = f(x, \lambda) \tag{6.2}$$

be a family of planar differential systems depending on one real parameter λ.

In Section 2.9 we saw all types of phase portrait for a planar linear autonomous system. Obviously, for nonlinear systems there are many more possibilities. According to a famous result, the *Poincaré*[2]*–Bendixson*[3] *theorem*, for planar systems the orbits can be equilibria, orbits approaching equilibria, orbits going to infinity, cycles (closed orbits corresponding to periodic solutions) and orbits that approach cycles. Note that

---

[1] Eberhard Frederich Ferdinand Hopf (1902–1983)
[2] Jules Henri Poincaré (1854–1912)
[3] Ivar Otto Bendixson (1861–1935)

this conclusion strictly holds for planar systems, and that in higher dimensions the orbital behavior of systems could be extremely complex.

For the next discussion, let us adopt the following terminology: by a stable equilibrium we shall mean an asymptotically stable equilibrium (stable focus), and by a *stable cycle* we shall understand a cycle (nontrivial periodic solution) such that all neighboring trajectories approach the cycle as $t \to +\infty$. Additionally, since any equilibrium $x(\lambda)$ of (6.2) by the substitution $y := x - x(\lambda)$, corresponds to the zero equilibrium of the transformed system $y' = f(y + x(\lambda), \lambda)$, and consequently any continuous branch of equilibrium points $x = x(\lambda)$ of the one-parameter family is reduced to the line $y = 0$, we may assume without loss of generality that 0 is an equilibrium of (6.2) for every λ, i.e. $f(0, \lambda) = 0$ for all λ.

The term Hopf bifurcation refers to the transition from the stability (instability) of the origin to the stability (instability) of cycles surrounding the origin as the parameter increasingly or decreasingly crosses a critical value. Therefore, at the bifurcation point, the origin loses the stability (instability) property, and gives birth to cycles surrounding it and taking over the stability (instability) property.

More exactly, the Hopf bifurcation occurs at $\lambda = \lambda_0$ if for some neighborhood U of λ_0, there is a continuous mapping $\varphi : \mathbb{R} \times U \to \mathbb{R}^2$ such that:
- $\varphi(\cdot, \lambda)$ is the zero solution of the system and the origin is a stable (unstable) equilibrium for all $\lambda < \lambda_0$, or for all $\lambda > \lambda_0$;
- the origin is an unstable (stable) equilibrium for all $\lambda > \lambda_0$ (respectively $\lambda < \lambda_0$);
- $\varphi(\cdot, \lambda)$ is a stable (respectively unstable) cycle surrounding the origin for all $\lambda > \lambda_0$ (respectively $\lambda < \lambda_0$).

The bifurcation is called *supercritical* if the bifurcating cycles are stable, and *subcritical* if they are unstable.

Hopf bifurcation typically occurs when the eigenvalues of the Jacobian matrix at the origin are complex conjugates and cross the imaginary axis; more exactly if:

(H). the eigenvalues of the Jacobian matrix $f_x(0, \lambda)$ in a neighborhood of λ_0 are $\alpha(\lambda) \pm i\beta(\lambda)$ with $\alpha(\lambda_0) = 0$, $\beta(\lambda_0) \neq 0$ and $\alpha'(\lambda_0) \neq 0$.

Then, if $\alpha'(\lambda_0) > 0$, one has that $\alpha(\lambda) < 0$ at the left of λ_0 and $\alpha(\lambda) > 0$ to the right of λ_0, so that the stability property of the origin changes from stability to instability at $\lambda = \lambda_0$. Additionally, if $\alpha'(\lambda_0) < 0$, then the origin goes from instability to stability.

Note however that condition (H) is not sufficient for the Hopf bifurcation. For example, condition (H) is satisfied for $\lambda_0 = 0$ in the case of the system

$$\begin{cases} x' = \lambda x + y \\ y' = -x + \lambda y, \end{cases}$$

for which the eigenvalues are $\lambda \pm i$. However, $\lambda = 0$ is not a Hopf bifurcation point since linear systems cannot have stable cycles.

Sufficient conditions for the Hopf bifurcation are known and can be found in the literature.

Example 6.5. Consider the equation $x'' - (\lambda - x^2)x' + x = 0$ and the equivalent system

$$\begin{cases} x' = y \\ y' = -x + (\lambda - x^2)y. \end{cases}$$

The origin is an equilibrium of the system for all λ. The eigenvalues of the Jacobian matrix at the origin are $(\lambda \pm i\sqrt{4-\lambda^2})/2$ for $\lambda \in (-2, 2)$. The origin is a stable equilibrium for $\lambda \in (-2, 0)$, and unstable equilibrium for $\lambda \in (0, 2)$. Additionally, for each small positive λ, there is a stable cycle (Figure I.6.1). Hence the system has a supercritical Hopf bifurcation at $\lambda = 0$.

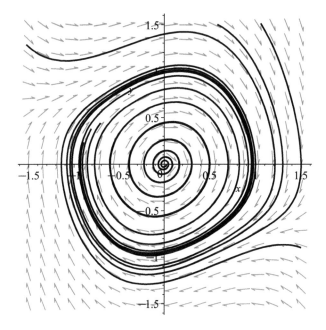

Fig. I.6.1: Example 6.5: phase portrait for $\lambda = 0.2$.

Example 6.6. Consider the system

$$\begin{cases} x' = -y + x(\lambda - x^2 - y^2) \\ y' = x + y(\lambda - x^2 - y^2). \end{cases}$$

By linearization, we can easily see that the origin is a stable equilibrium for $\lambda < 0$. To analyze the orbits for $\lambda > 0$, let us convert the system to polar coordinates by making $x = r\cos\theta$ and $y = r\sin\theta$. Substituting into the equations, multiplying the first equation by $\cos\theta$ and the second one by $\sin\theta$, and then adding them, gives $r' = r(\lambda - r^2)$.

Additionally, multiplying the first equation by $\sin\theta$ and the second one by $-\cos\theta$, and then adding them, gives $\theta' = 1$. Thus we arrive at the system

$$\begin{cases} r' = r(\lambda - r^2) \\ \theta' = 1. \end{cases}$$

Hence $\theta = t + c$, where c is an arbitrary constant. This shows that the motion on orbits is counterclockwise. Now we solve the first equation under the initial condition $r(0) = r_0$. To this aim, it is convenient to rewrite the equation as $rr' = r^2(\lambda - r^2)$, or $(r^2)' = 2r^2(\lambda - r^2)$. Next we set $s = r^2$ and separate variables to finally obtain

$$r^2 = \frac{\lambda r_0^2}{r_0^2 + (\lambda - r_0^2)e^{-2\lambda t}}.$$

This formula shows that any orbit starting from a point located inside the circle of radius $\sqrt{\lambda}$ and centered at the origin, i.e. with $r_0 < \sqrt{\lambda}$, will remain inside the circle, and steadily approach the circle as $t \to +\infty$. If the orbit starts from a point located outside the circle, i.e. if $r_0 > \sqrt{\lambda}$, then it will stay outside and steadily approach the circle as $t \to +\infty$. Hence the circle $r = \sqrt{\lambda}$, itself an orbit of the system, is approached by the orbits from its interior and its exterior, that is, it is a stable cycle. Therefore the system has a supercritical Hopf bifurcation at $\lambda = 0$.

Example 6.7. Consider the system

$$\begin{cases} x' = -y - x(\lambda - x^2 - y^2) \\ y' = x - y(\lambda - x^2 - y^2). \end{cases}$$

In polar coordinates one has

$$\begin{cases} r' = r(-\lambda + r^2) \\ \theta' = 1. \end{cases}$$

Solving the first equation under the initial condition $r(0) = r_0$ yields

$$r^2 = \frac{\lambda r_0^2}{r_0^2 + (\lambda - r_0^2)e^{2\lambda t}}.$$

This formula shows that for $\lambda > 0$, any orbit starting from a point located inside the circle of radius $\sqrt{\lambda}$, and centered at the origin, will remain inside the circle, and approach the origin as $t \to +\infty$. Additionally, if the orbit starts from a point located outside the circle, then the corresponding solution blows up in finite time, as $t \to t_0$, where $r_0^2 + (\lambda - r_0^2)e^{2\lambda t_0} = 0$. So, the orbit moves away from the cycle $r = \sqrt{\lambda}$, which is now unstable. The origin is also an unstable equilibrium for $\lambda < 0$. Therefore the system has a subcritical Hopf bifurcation at $\lambda = 0$.

6.3 Optimization of Differential Systems

At the beginning of this chapter, we called control parameters those parameters of a differential system that are varied, contrary to the parameters that are assumed to be fixed. Then an analysis of the equilibria was carried out with respect to the control parameters, as number of equilibria and stability type. The 'control' role of the variable parameters comes when we are interested in 'optimizing' the differential system by choosing those values of the parameters that lead to a suitable behavior of the system. More often the optimization of a differential system is related to the minimization or maximization of some functional associated with the system. To be more explicit, let us consider some examples.

Example 6.8 (Optimal one-species harvesting). In the case of the already considered equation

$$x' = ax\left(1 - \frac{x}{k}\right) - ux$$

modeling the dynamics of an animal (fish or bird) population in the presence of harvesting, we may be interested in the condition where the harvesting does not lead to the extinction of the population. In this case, the control parameter u, expressing the intensity of harvesting per capita, should be less than the critical value $u_c = a$, as shown by Example 6.1. Then the estimate of the population level is of $(a-u)k/a$. If we wish to guarantee a lower bound x_L of the population, then from the inequality

$$\frac{(a-u)k}{a} \geq x_L \tag{6.3}$$

we obtain the admissible values of the control parameter, $u \leq u_*$, where $u_* = (k - x_L)a/k$.

Furthermore, let us assume that both coefficients u and k are variable, and that there is a profit of harvesting depending on the effort coefficient u and on the carrying capacity k. Let the profit be given by the function $\varphi(u, k)$. Then we may put the problem of finding u and k for which the profit is maximal under the constraint condition (6.3) and some a priori bounds for u and k, $u \geq u_L$ and $0 < k \leq k_L$, with

$$\frac{(a - u_L)k_L}{a} > x_L, \quad u_L < \frac{a}{2} \quad \text{and} \quad x_L < \frac{k_L}{2}.$$

Mathematically we arrive at the constraint optimization problem

$$\max_{u,k} \varphi(u, k), \quad \text{subject to} \tag{6.4}$$

$$(a - u)k \geq ax_L, \quad u \geq u_L, \quad 0 < k \leq k_L.$$

For instance, we may consider that the profit is jointly proportional to the harvesting effort and the inverse of the carrying capacity, i.e.

$$\varphi(u, k) = \frac{cu}{k}.$$

This means that the profit increases proportionally with the total catch uk from the equilibrium population, and decreases with the square of the carrying capacity due to the food costs. Of course, the proportionality coefficient c depends on the existing population x_0 at the precise time t_0 when the optimization problem is stated, that is $c = c(x_0)$. If we denote $u_1 := k$, $u_2 = uk$, then we have a nonlinear optimization problem with linear constraints,

$$\max_{u_1,u_2} c\frac{u_2}{u_1^2}, \quad \text{subject to}$$

$$au_1 - u_2 \geq ax_L, \quad u_2 - u_L u_1 \geq 0, \quad 0 < u_1 \leq k_L.$$

This can be easily solved geometrically if we represent the admissible domain of the problem in the $u_1 O u_2$ plane, i.e. the set of points that satisfy the constraints. Furthermore, we see on the resulting graphic representation that the maximal profit μ_{max} is the largest positive μ for which the parabola $cu_2/u_1^2 = \mu$ intersects the admissible domain, and this happens for that parabola which is tangent to the line $au_1 - u_2 = ax_L$. The tangency condition, namely the condition that the curves $cu_2/u_1^2 = \mu$ and $au_1 - u_2 = ax_L$ have a unique intersection point, gives, after some calculation, the maximal profit

$$\mu_{max} = \frac{ac}{4x_L},$$

which is reached for the optimal values

$$u_1^o = 2x_L, \quad u_2^o = ax_L.$$

These give the values of the control coefficients for an optimal one-species harvesting policy,

$$k_o = 2x_L, \quad u_o = \frac{a}{2}.$$

Example 6.9 (Optimal two-species harvesting). Consider the Lotka–Volterra model with harvesting

$$\begin{cases} x' = x(r - ay) - u_1 x \\ y' = -y(m - bx) - u_2 y, \end{cases}$$

where the coefficients u_1, u_2 give a measure of the capture effort for each species. The equilibria of the system are $(0, 0)$ and (x^*, y^*), where $x^* = (m + u_2)/b$ and $y^* = (r - u_1)/a$. The stability analysis yields to the conclusion that the zero solution is asymptotically stable for $u_1 > r$, while for $u_1 < r$, (x^*, y^*) is stable. Hence, to avoid extinction of the species one has to guarantee that $u_1 < r$. Assume now that the harvesting profit is given by

$$\varphi(u_1, u_2) = c_1 u_1 x^* + c_2 u_2 y^*$$

and that the wish is to maximize the profit guaranteeing that the ratio x^*/y^* remains between two given numbers α and β satisfying $\alpha < ma/(br) < \beta$, i.e.

$$\alpha \leq \frac{m + u_2}{b} \cdot \frac{a}{r - u_1} \leq \beta.$$

Hence we obtain the optimization problem

$$\max_{u_1,u_2} \left(\frac{c_1}{b}(m+u_2)u_1 + \frac{c_2}{a}(r-u_1)u_2 \right), \quad \text{subject to}$$

$$ab(r-u_1) \le a(m+u_2) \le \beta b(r-u_1).$$

Example 6.10 (Optimal treatment of leukemia). Consider the hematologic model already mentioned in Section 4.1

$$\begin{cases} x' = \frac{ax}{1+b(x+y)} - cx \\ y' = \frac{Ay}{1+B(x+y)} - Cy. \end{cases}$$

Here $x(t)$ and $y(t)$ represent the number of normal and leukemic cells respectively at time t; a, A are the growth rates of the two cell populations; c, C are the cell death rates; and b, B stand for the sensibility of the two types of cells with respect to the crowding effect in the bone marrow microenvironment. The parameters a, b and c are universal in all human beings, while the parameters A, B and C may differ from one person to another. The equilibria of the system are obtained by solving the algebraic system

$$\frac{ax}{1+b(x+y)} - cx = 0, \quad \frac{Ay}{1+B(x+y)} - Cy = 0.$$

They are $(0,0)$, $(d,0)$ and $(0,D)$, where

$$d = \frac{1}{b}\left(\frac{a}{c} - 1\right), \quad D = \frac{1}{B}\left(\frac{A}{C} - 1\right).$$

Using the linearization method, we can show that the solution $(0,0)$ is always unstable. Also, from the solutions $(d,0)$ and $(0,D)$, only one is asymptotically stable, namely $(d,0)$ if $D < d$, and $(0,D)$ if $D > d$. Hence, in the case where $D < d$, the normal cell population $x(t)$ approaches the equilibrium abundance d, while the leukemic cell population $y(t)$ tends to zero. In contrast, if $D > d$, then $(0,D)$ is asymptotically stable, that is the leukemic cell population becomes dominant approaching its equilibrium abundance D and leads in the limit to the elimination of the normal cells, i.e. $x(t)$ goes to zero. Thus, we may say that from a hematologic point of view, the inequality $D < d$ characterizes healthy people, while the converse inequality $D > d$ characterizes the people with leukemia. Mathematically, in terms of the parameter D, the value d is a bifurcation point of the system, corresponding to the transition from the normal hematologic state to the leukemic state.

Based on the above conclusions, to eradicate the cancer a targeted therapy should act on the parameters A, B and C, in order to reverse the inequality $D > d$. This is possible by decreasing the growth rate A and/or by increasing the sensibility rate B and cell death rate C. Thus, the treatment should change the values A, B, C with $u_1 A, u_2 B$ and $u_3 C$, where $0 < u_1 \le 1$, $u_2 \ge 1$, $u_3 \ge 1$, in such a way that $D(u_1, u_2, u_3) < d$, where

$$D(u_1, u_2, u_3) := \frac{1}{u_2 B}\left(\frac{u_1 A}{u_3 C} - 1\right).$$

Correspondingly, we have the following system:

$$\begin{cases} x' = \dfrac{ax}{1+b(x+y)} - cx \\ y' = \dfrac{u_1 A y}{1+u_2 B(x+y)} - u_3 C y, \end{cases}$$

with control parameters u_1, u_2, u_3. Obviously, there is a cost $\varphi(u_1, u_2, u_3)$ of the treatment and the interest to minimize it. Therefore, the optimal treatment corresponds to the solution of the optimization problem

$$\min_{u_1, u_2, u_3} \varphi(u_1, u_2, u_3), \quad \text{subject to}$$

$$\frac{1}{u_2 B}\left(\frac{u_1 A}{u_3 C} - 1\right) < d, \quad 0 < u_1 \le 1, \quad u_2 \ge 1, \quad u_3 \ge 1.$$

To give an example of a cost function, assume that the costs are known for one per cent modification of the parameters A, B, C; let them be c_1, c_2, c_3 respectively. Then for the reduction from A to $u_1 A$ of the growth rate of leukemic cells, the cost will be $c_1(1-u_1)$. Similarly, for the increase from B to $u_2 B$, and from C to $u_3 C$ of the sensibility and cell death rates, the costs will be $c_2(u_2 - 1)$ and $c_3(u_3 - 1)$ respectively. Hence the total cost is

$$\varphi(u_1, u_2, u_3) = c_1(1 - u_1) + c_2(u_2 - 1) + c_3(u_3 - 1).$$

Note that the general *constraint optimization problem* is

$$\min_u (\max_u) \varphi(u), \quad \text{subject to}$$

$$\chi_i(u) = c_i \quad \text{for } i = 1, 2, \ldots, p,$$

$$\psi_j(u) \ge d_j \quad \text{for } j = 1, 2, \ldots, q,$$

where $u = (u_1, u_2, \ldots, u_n)$. A good introduction to Optimization Theory is the book V. G. Karmanov, *Mathematical Programming*, Mir Publishers, Moscow, 1989.

6.4 Dynamic Optimization of Differential Systems

Although the models in the previous section deal with dynamic processes, their parameters have been assumed not to be time dependent. Consequently, the optimal choice of the control parameters has been given by the solution of a *static optimization* problem. In many cases, a more realistic approach is to let the control parameters be time dependent and to optimize processes over time. In these cases, one speaks about *dynamic optimization* or *optimal control*. It can be in *discrete time* or in *continuous time*.

For an example, let us come back to Example 6.8 and try to find the optimal harvesting policy over a time interval $[0, T]$. To this aim, we divide the interval $[0, T]$ into n small intervals $[t_{i-1}, t_i]$, where $0 = t_0 < t_1 < \ldots < t_n = T$, and we assume that for each subinterval $[t_{i-1}, t_i]$, the population density and the harvesting effort are constant; let them be x_i and u_i. Also assume that the harvesting benefit during this time interval is of the form $(t_i - t_{i-1})f(t_i, x_i, u_i)$. Hence the total profit over the period $[0, T]$ is

$$\varphi = \sum_{i=1}^{n}(t_i - t_{i-1})f(t_i, x_i, u_i).$$

As for the values x_i we may assume a transition relation coming from the discretization of the differential equation, namely the difference equation

$$\frac{x_i - x_{i-1}}{t_i - t_{i-1}} = ax_{i-1}\left(1 - \frac{x_{i-1}}{k}\right) - u_{i-1}x_{i-1}, \quad i = 1, 2, \ldots, n,$$

where the initial population density x_0 is assumed to be known. Thus we obtain the optimization problem

$$\max_{u_1, u_2, \ldots, u_n} \sum_{i=1}^{n}(t_i - t_{i-1})f(t_i, x_i, u_i), \quad \text{subject to} \quad (6.5)$$

$$x_i = x_{i-1} + (t_i - t_{i-1})\left[ax_{i-1}\left(1 - \frac{x_{i-1}}{k}\right) - u_{i-1}x_{i-1}\right].$$

Some other constraints, as above, on x_i and u_i are also added. A typical example is a fish population that is harvested for its commercial value. Then a reasonable expression of the net profit is

$$f(t_i, x_i, u_i) = e^{-\delta t_i}(c_1 u_i x_i - c_2(u_i x_i)^2 - c_3 u_i),$$

where $c_1 u_i x_i$ is the profit from the sale of fish, $c_2(u_i x_i)^2$ represents the diminishing returns due to market saturation, $c_3 u_i$ is the cost of fishing, and $e^{-\delta t_i}$ is a discount factor.

Problem (6.5) is the discrete-time version of the optimal control problem in continuous time

$$\max_{u} \int_{0}^{T} f(t, x(t), u(t))\, dt, \quad \text{subject to}$$

$$x' = ax\left(1 - \frac{x}{k}\right) - u(t)x, \quad x(0) = x_0.$$

More generally, the optimal control problem associated with a first-order differential system can be stated as follows:

$$\max_{u} \int_{0}^{T} f(t, x(t), u(t))\, dt, \quad \text{subject to}$$

$$x' = g(t, x, u), \quad x(0) = x_0,$$

where some other constraints given by equalities and inequalities and involving both *state variable* x and *control variable* u are usually added. A solution $x^*(t), u^*(t)$ of the problem is called an *optimal pair*.

For example, the fishery model leads to the control problem

$$\max_u \int_0^T e^{-\delta t}(c_1 ux - c_2(ux)^2 - c_3 u)\, dt, \quad \text{subject to}$$

$$x' = ax\left(a - \frac{x}{k}\right) - u(t)x, \quad x(0) = x_0.$$

The basic problems in optimal control theory are: the appropriate formulation of the control problem; the existence and the uniqueness of the optimal control; the numerical computation of the optimal control; and its dependence on various parameters of the system. A good introduction to optimal control theory is the book by B. D. Craven [6].

Part II: **Exercises**

Seminar 1
Classes of First-Order Differential Equations

1.1 Solved Exercises

Exercise 1.1. Solve the equation

$$(t^2 + 1)x' = 2tx$$

and find the solution of the Cauchy problem with the initial condition $x(0) = 2$.

Solution. The equation has separable variables. Separating variables gives

$$\frac{x'}{x} = \frac{2t}{t^2 + 1}.$$

Integration with respect to t yields $\ln|x| = \ln(t^2 + 1) + \ln C$, where C is an arbitrary positive constant. This gives the general solution $x = C(t^2 + 1)$, $C \in \mathbb{R}$. To solve the Cauchy problem, we let $t = 0$. Then $x(0) = C$, and so $C = 2$. Therefore, the solution of the Cauchy problem is $x = 2(t^2 + 1)$.

Exercise 1.2. Solve the equation

$$t^2 x' = x^2 + tx + t^2$$

for $t > 0$, and the Cauchy problem with the initial condition $x(1) = 0$.

Solution. The equation can be written in the form

$$x' = \frac{x^2 + tx + t^2}{t^2}.$$

The right-hand side is a rational function in t and x, whose numerator and denominator are homogeneous polynomials having the same degree of homogeneity, in this case 2. Thus the equation is of homogeneous type. It can be put in the form

$$x' = \left(\frac{x}{t}\right)^2 + \frac{x}{t} + 1.$$

Making the substitution $y = x/t$ yields $y + ty' = y^2 + y + 1$, or $ty' = y^2 + 1$. Next, separate the variables,

$$\frac{y'}{y^2 + 1} = \frac{1}{t},$$

where by integration $\arctan y = \ln t + C$. Hence $y = \tan(\ln t + C)$, and finally the general solution is $x = t\tan(\ln t + C)$, $C \in \mathbb{R}$. For $t = 1$, one has $\tan C = 0$, hence $C = k\pi$, $k \in \mathbb{Z}$. Therefore, the solution of the Cauchy problem is $x = t\tan(\ln t)$.

Exercise 1.3. Find the solutions of the equation
$$x' = \frac{x+t-3}{x-t+1}.$$

Solution. If the numerator and denominator were homogeneous polynomials in x, t, then, as above, the equation would be of homogeneous type. So the idea is to make suitable changes of variables that lead to the above-mentioned situation. Try the changes of variables $s := t - a$, $y := x - b$. Then $x + t - 3 = y + b + s + a - 3$ and $x - t + 1 = y + b - s - a + 1$. If we require that the constant terms are zero, that is
$$\begin{cases} a + b - 3 = 0 \\ -a + b + 1 = 0, \end{cases}$$
we obtain $a = 2$, $b = 1$. In the new variables s and y, the differential equation reads as
$$y' = \frac{y+s}{y-s},$$
or equivalently
$$y' = \frac{\frac{y}{s}+1}{\frac{y}{s}-1}.$$

Thus by suitable changes of variables the equation is reduced to an equation of homogeneous type. Let $z := y/s$. Then $y = sz$, $y' = z + sz'$ and
$$z + sz' = \frac{z+1}{z-1}.$$

It follows that
$$sz' = -\frac{z^2 - 2z - 1}{z - 1},$$
or, after separating the variables,
$$\frac{z-1}{z^2 - 2z - 1} z' = -\frac{1}{s}.$$

Integrating in s yields $\ln|z^2 - 2z - 1| = -2\ln|s| + \ln C$. Hence $z^2 - 2z - 1 = C/s^2$. Since $z = y/s = (x-1)/(t-2)$, we obtain the general solution in implicit form: $(x-1)^2 - 2(t-2)(x-1) - (t-2)^2 = C$.

Exercise 1.4. Solve the differential equation
$$xy' = y - 1$$
for the unknown $y = y(x)$.

Solution. The equation is linear. First we solve the associated homogeneous equation $xy' = y$. Its general solution is $y_h = Cx$, $C \in \mathbb{R}$. Next we look for a particular solution of the nonhomogeneous equation, of the form $y_p = C(x)x$. Substituting this into the equation gives $C'(x)x^2 = -1$. Hence $C'(x) = -1/x^2$, where $C(x) = 1/x$. Therefore, $y_p = 1$, and the general solution is $y = Cx + 1$.

Exercise 1.5. Solve the following equation for $y = y(x)$:

$$y' = y + xy^2 .$$

Solution. The equation is of Bernoulli type. Division by y^2 yields $y'y^{-2} = y^{-1} + x$. Next we change the variable via $z := y^{-1}$, and obtain the linear equation $z' = -z - x$. Its general solution is $z = Ce^{-x} + 1 - x$. Finally, the general solution of the original equation is $y = (Ce^{-x} + 1 - x)^{-1}$, $C \in \mathbb{R}$.

Exercise 1.6. Reduce the following equation to a Bernoulli equation:

$$tx' = -x + t^3 x^2 - 4t ,$$

if we know that it has solutions of the form $x = \lambda/t$.

Solution. The equation is of Riccati type. Substituting $x = \lambda/t$ into the equation gives $\lambda^2 = 4$. Choose $\lambda = 2$ and let $y := x - 2/t$. Then $x = y + 2/t$, $x' = y' - 2/t^2$ and

$$ty' - \frac{2}{t} = -y - \frac{2}{t} + t^3 \left(y + \frac{2}{t}\right)^2 - 4t .$$

After some calculation one obtains the Bernoulli equation

$$y' = -\frac{y}{t} + t^2 y^2 + 4ty .$$

1.2 Proposed Exercises

Exercise 1.7. Solve the following differential equations with separable variables:

(a) $x^3 y' + \sqrt{1 - y^2} = 0$;
(b) $\sqrt{x} y' = \sqrt{1 + y^2}$;
(c) $(t^2 - 1)u' = u^2 - 1$.

Exercise 1.8. Find the general solution of the following equations of homogeneous type:

(a) $t^3 x' + x^3 = t^2 x$ $(t > 0)$;
(b) $2xyy' = x^2 + y^2$ $(x > 0)$;
(c) $xy' = y - 2\sqrt{xy}$ $(x > 0)$.

Exercise 1.9. Consider the differential equation

$$x' = \frac{\lambda x - 3t - 1}{x - t + 1} ,$$

where λ is a real parameter.
(a) Show that for $\lambda \neq 3$, the equation can be reduced to an equation of homogeneous type;
(b) Solve the equation for $\lambda = 3$ by making the substitution $y = x - t$.

Exercise 1.10. Solve the following linear equations:

(a) $tx' + x - e^{-t} = 0 \quad (t > 0)$;
(b) $x^2 y' = y + x^2 e^{-\frac{1}{x}} \quad (x > 0)$;
(c) $u' = e^{2x} - u$.

Exercise 1.11. Find the solutions of the following Bernoulli equations:

(a) $x' = \frac{x}{t} + 2\sqrt{x} \quad (t > 0)$;
(b) $x' = \frac{2}{t}x + \frac{1}{x} \quad (t > 0)$;
(c) $y' = y - xy^3$.

1.3 Solutions

Exercise 1.7
(a) $y = \sin\left(\frac{1}{2x^2} + C\right)$, $C \in \mathbb{R}$.
(b) $y = \sinh(2\sqrt{x} + C)$, $C \in \mathbb{R}$.
(c) $u = \frac{(1+C)t+1-C}{(1-C)t+1+C}$, $C \in \mathbb{R}$.

Exercise 1.8
(a) $x = \pm \frac{t}{\sqrt{2\ln t + C}}$, $C \in \mathbb{R}$.
(b) $y = \pm\sqrt{Cx + x^2}$, $C \in \mathbb{R}$.
(c) $y = x(C - \ln x)^2$, $C \in \mathbb{R}$.

Exercise 1.9
(a) The changes of variables are $y = x - a$, $s = t - b$, where $a = 4(\lambda - 3)^{-1}$, $b = (\lambda + 1)(\lambda - 3)^{-1}$;
(b) $x - 3t + 2\ln|x - t - 1| = C$ (in the implicit form);

Exercise 1.10
(a) $x = \frac{C - e^{-t}}{t}$;
(b) $y = (x + C)e^{-\frac{1}{x}}$;
(c) $u = Ce^{-x} + \frac{1}{3}e^{2x}$.

Exercise 1.11
(a) $\sqrt{x} = 2t + C\sqrt{t}$;
(b) $x = \pm\frac{1}{3}(9Ct^4 - 6t)^{1/2}$;
(c) $y = \pm\left(x - \frac{1}{2} + Ce^{-2x}\right)^{-1/2}$.

1.4 Project: Problems of Geometry that Lead to Differential Equations

Suggested structure. 1. The differential equation of a family of planar curves; 2. The differential equation of orthogonal trajectories of a family of curves; 3. Other problems of geometry leading to differential equations; and 4. Lagrange[1] and Clairaut[2] differential equations (examples of geometrical problems leading to such equations; their solving).

Basic bibliography. B. Demidovich (ed.), *Problems in Mathematical Analysis*, Mir Publishers, Moscow (Chapter 9); C. C. Ross [23, pp. 90–95].

Hints. 1. Consider the family of planar curves $y = f(x, C)$ depending on a real parameter C, where f is assumed to be of class C^1 with respect to the first variable. Differentiating with respect to x yields $y' = f'(x, C)$. The elimination of C from the two relations leads to a differential equation of the form $F(x, y, y') = 0$, called the *differential equation of the family of curves*.

Example 1a: Consider the family of parabolas $y = Cx^2$, $C \in \mathbb{R}$. Then $y' = 2Cx$. The elimination of the parameter yields the differential equation

$$y' = \frac{2y}{x}.$$

Conversely, the general solution of this equation with separable variables is the family of curves $y = Cx^2$, $C \in \mathbb{R}$.

Example 1b: If a family of curves is given by an equation that is not explicit in y, then we consider y as a function of x, i.e. $y = y(x)$, and we differentiate the equation of the curve family with respect to x. For instance, for the family of ellipses $x^2 + Cy^2 = 1$ ($C > 0$), by differentiating we obtain $2x + 2Cyy' = 0$, and elimination of the parameter C yields the differential equation

$$y' = \frac{xy}{x^2 - 1}.$$

2. *Orthogonal trajectories* are a family of curves in the plane that intersect a given family of curves at right angles. If the differential equation of the given family of curves is $F(x, y, y') = 0$, then

$$F\left(x, y, -\frac{1}{y'}\right) = 0$$

is the *differential equation of the orthogonal trajectories*.

---

[1] Joseph-Louis Lagrange (1736–1813)
[2] Alexis Claude de Clairaut (1713–1765)

Example 2: The differential equation of the orthogonal trajectories of the family of parabolas $y = Cx^2$, $C \in \mathbb{R}$ is
$$-\frac{1}{y'} = \frac{2y}{x}.$$
Separating variables gives $2yy' = -x$, where $y^2 = -x^2/2 + C$. Therefore, the orthogonal trajectories are the ellipses $x^2 + 2y^2 = C$.

3. Example 3a: Find all the curves $y = y(x)$ ($x > 0$) such that the y-intercept of the tangent to the curve at an arbitrary point is equal to the abscissa of the tangency point. The equation of the tangent line at $(x_0, y(x_0))$ is $y = y(x_0) + y'(x_0)(x - x_0)$. Its y-intercept (y-coordinate for $x = 0$) is $y(x_0) - x_0 y'(x_0)$. Hence the required condition becomes $y(x_0) - x_0 y'(x_0) = x_0$. As the point x_0 is arbitrary, one obtains the (linear) differential equation $y - xy' = x$. Its general solution is $y = Cx - x \ln x$, $C \in \mathbb{R}$.

Example 3b: Find all the curves $y = y(x)$ ($x > 0$) such that the product of the y-intercept of the tangent to the curve at an arbitrary point and the abscissa of the tangency point is a given constant a.

4. Lagrange and Clairaut differential equations are first-order differential equations expressing y with x and y'. *Lagrange's differential equation* has the form
$$y = x\phi(y') + \psi(y'),$$
where ϕ, ψ are given functions of class C^1, and $\phi(p) \neq p$. For $\phi(p) \equiv p$, one has *Clairaut's differential equation*
$$y = xy' + \psi(y').$$
Use the bibliography (e.g. I. A. Rus [24, pp. 46–47] and I. I. Vrabie [26, pp. 25–27]) to learn how to solve the Lagrange and Clairaut equations.

Seminar 2
Mathematical Modeling with Differential Equations

2.1 Solved Exercises

Exercise 2.1 (Tumor growth). Assume that the growth rate of a spherical tumor is proportional to its volume. Write down and solve the differential equation modeling the tumor growth. Then, assuming that the tumor doubles its volume in 100 days, find the tumor radius as function of time, and the time taken by the tumor to double its radius.

Solution. Let $v(t)$ be the tumor volume at time t (days). Then $v' = av$, where a is a proportionality factor depending on the aggressiveness of cancer cells. Solving the equation yields the growth law
$$v(t) = e^{at}v_0, \qquad (2.1)$$
where v_0 is the initial tumor volume at time $t_0 = 0$.

Assuming that the tumor doubles its volume in 100 days, one has $2v_0 = e^{100a}v_0$. Hence $100a = \ln 2$, that is $a = 10^{-2}\ln 2$. Let $r(t)$ be the tumor radius at time t. Then (2.1) becomes $4\pi r(t)^3/3 = e^{at}4\pi r_0^3/3$, and the growth law of the tumor radius is
$$r(t) = 2^{\frac{10^{-2}t}{3}}r_0.$$
Finally if the radius doubles, then $2r_0 = 2^{10^{-2}t/3}r_0$, so that $t = 300$.

Exercise 2.2 (Rumor spread). A rumor spreads through a fixed population of N people. Find the spread law assuming that the spread rate is one of the following cases: (a) proportional to the population that has heard the rumor; (b) proportional to the population that has not heard the rumor; (c) jointly proportional to the number of people who have and have not heard the rumor. In each case determine the spread of the rumor over a long time (as time goes to infinity) and summarize the appropriateness of the corresponding model.

Solution. Denote by $x(t)$ the number of people from the community who have heard the rumor by time t, and by x_0 the number of people who have started to spread the rumor, that is $x_0 = x(0)$.

In case (a), the model is given by the Malthus equation $x' = ax$, where $a > 0$ is a proportionality factor depending on the appetence for rumors of that community, on the interest on a particular rumor and on the rapidity of communication within the community. The solution satisfying the initial condition $x(0) = x_0$ is
$$x(t) = e^{at}x_0$$
and represents the spread law of the rumor inside the given community. To use this law, it is necessary to determine the constant a. This can be done if we know the number x_1 of the people who have heard the rumor at another moment $t_1 > 0$. Then, for

$t = t_1$, one has $x_1 = e^{at_1}x_0$, so that $at_1 = \ln(x_1/x_0)$. Hence

$$a = \frac{1}{t_1} \ln \frac{x_1}{x_0}.$$

In case (b), the model is $x' = a(N - x)$. The solution satisfying the initial condition $x(0) = x_0$ is

$$x(t) = N - e^{-at}(N - x_0).$$

As above, one can find the value of the proportionality constant a,

$$a = \frac{1}{t_1} \ln \frac{N - x_0}{N - x_1}.$$

In case (c), the model is given by the logistic equation $x' = ax(N - x)$. Then the spread law of the rumor is

$$x(t) = \frac{N}{1 - \left(1 - \frac{N}{x_0}\right)e^{-aNt}}.$$

Long-term behavior of solutions: (a) $\lim_{t \to +\infty} x(t) = +\infty$, although we must have $x(t) \leq N$. Therefore, the first model is only useful for a short period of time; (b) $\lim_{t \to +\infty} x(t) = N$, showing that the second model is appropriate; (c) $\lim_{t \to +\infty} x(t) = N$, so the third model is also believable.

Exercise 2.3 (Innovation adoption). A new innovation (e.g. a new communication technology, a new industrial technology, a new drug for the treatment of a disease, a new agricultural pesticide, etc.) is introduced at time $t = 0$ in a community of N potential users. Denote by $x(t)$ the number of users who have adopted the new innovation by time t. Assume that the rate of adoption of the new innovation is jointly proportional to the number of users who have adopted it and the number of users who have not adopted it. Write down the corresponding mathematical model and analyze how $x(t)$ behaves over time.

Solution. As in case (c) from the previous problem, the model is given by the logistic equation $x' = ax(N - x)$, and the conclusion is that there is an increasing trend of adoption, which tends over time to equal the total number of users.

Exercise 2.4 (A criminalistic problem). The body of a victim was discovered at time t_1 when its temperature was T_1. The temperature of the body was taken once more later at the crime scene, at time t_2 ($t_2 > t_1$) and found to be T_2. Assuming that the temperature of the environment was constant T_e, and knowing the temperature T_0 of the alive human body, find victim's time of death.

Solution. Let $T(t)$ be the victim's body temperature at time t. One has $T(t_0) = T_0$. We use Newton's law of heat transfer, $T' = a(T_e - T)$. Solving the equation yields the changing law of the victim's body temperature

$$T(t) = T_e + e^{-a(t-t_0)}(T_0 - T_e).$$

Letting successively $t = t_1$ and $t = t_2$, gives a system of two equations in the unknowns a and t_0,

$$\begin{cases} T_e + e^{-a(t_1-t_0)}(T_0 - T_e) = T_1 \\ T_e + e^{-a(t_2-t_0)}(T_0 - T_e) = T_2. \end{cases}$$

This implies

$$\begin{cases} -a(t_1 - t_0) = \ln \frac{T_1-T_e}{T_0-T_e} \\ -a(t_2 - t_0) = \ln \frac{T_2-T_e}{T_0-T_e}. \end{cases}$$

Next by division we eliminate the parameter a and find the equation in the unknown t_0:

$$\frac{t_1 - t_0}{t_2 - t_0} = \frac{\ln|T_1 - T_e| - \ln|T_0 - T_e|}{\ln|T_2 - T_e| - \ln|T_0 - T_e|}.$$

Solving for t_0 gives the victim's time of death

$$t_0 = \frac{t_1 \ln|T_2 - T_e| - t_2 \ln|T_1 - T_e| + (t_2 - t_1)\ln|T_0 - T_e|}{\ln|T_2 - T_e| - \ln|T_1 - T_e|}.$$

2.2 Proposed Exercises

Exercise 2.5. Under normal conditions, the *E. coli* bacterium divides every 20 minutes. After 48 hours how many *E. coli* bacteria are there, if we start with a single bacterium?

Exercise 2.6. If the volume doubling time of a tumor is 60 days, after how many days does the tumor reach three times its initial volume?

Exercise 2.7. The half-life for radium is 1600 years. What is the percentage of the disintegrated substance after 100 years?

Exercise 2.8. The temperature inside a house is $x_0 = 20\,°C$ when the heating system fails. Assume that the exterior temperature decreases linearly during the next month (30 days) from $18\,°C$ to $8\,°C$, and that the heat transfer coefficient is $k = 3$. Find the formula that expresses the temperature variation in the house in the course of the month that follows the heating system failure.

Exercise 2.9. During cold winters the thickness of ice on a lake grows with a speed inversely proportional to the thickness. Initially, the ice was 2 cm thick and after 4 hours, 3 cm thick. How thick will the ice be after 6 more hours?

Exercise 2.10 (Resistance to reform). A reform or change is introduced in a system (an educational system, a health system, a judicial system, etc.) with N units, first at x_0 units. The implementation rate is directly proportional to the number of units that have already adopted change and inversely proportional to those that have not adopted it yet out of inertia. Write down the model of this process and prove that, no matter the resistance, the reform will impose itself. Also, prove that the stronger the resistance to reform is, the longer it takes for all units to adopt it.

2.3 Hints and Answers

Exercise 2.5
Use Malthus's model; $N = 2^{144}$.

Exercise 2.6
After $60 \ln 3 / \ln 2 \simeq 95$ days.

Exercise 2.7
$(1 - 2^{-1/16})100\% \simeq 4.2\%$.

Exercise 2.8
$x' = -k(x - x_e)$ (a nonhomogeneous linear equation), where $x_e = 18 - t/3$. Solving with the initial condition $x(0) = 20$, gives $x(t) = 17e^{-3t}/9 - t/3 + 163/9$.

Exercise 2.9
The ice will be 4 cm thick.

Exercise 2.10
The model is $x' = ax/(N - x)$. Notice that the stronger the resistance to reform is, the smaller the parameter $a > 0$ is. The equation has separable variables and its solution satisfying the initial condition $x(0) = x_0$ is obtained in the implicit form,

$$x - x_0 - N(\ln x - \ln x_0) = -at.$$

The reform is complete when $x = N$, and this happens at time

$$t(a) = \frac{N(\ln N - \ln x_0) - N + x_0}{a}.$$

One can show that $t(a)$ is positive and hence $t(a) \to +\infty$ as $a \to 0^+$.

2.4 Project: Influence of External Actions over the Evolution of Some Processes

Suggested structure. 1. The influence of advertising in marketing problems; 2. The influence of harvesting and predators in ecological problems; 3. Nonlinear diminishing of growth rates.

Basic bibliography. D. Kaplan and L. Glass [13], J. D. Logan [15].

2.4 Project: Influence of External Actions over the Evolution of Some Processes

Hints. 1. Let $S(t)$ be the number of sales from month t of a given product. Assume that the interest for that product will decrease over time and the sales diminishing rate is proportional to the volume. Therefore $S' = -aS$, where $a > 0$ is a proportionality constant. In order to sustain sales, one makes use of advertising. Its effect is an increase of sales, thus the equation becomes $S' = -aS + R$, where R is the advertising contribution to the sales growth rate. If we assume that due to the finite number of potential customers there is an upper limit of sales, let it be M, and that the advertising effect is directly proportional to the number of potential customers, then we may consider that R has the form $br(t)(M - S)/M$. Here $r(t)$ represents the intensity of advertising, which is not necessarily constant, and $(M - S)/M$ is the proportion of that part of the market that has still not purchased the product. Finally, by the proportionality constant $b > 0$, the advertising impact is taken into account for a given product and a given category of customers. Therefore, a mathematical model for the sales response to advertising ([15, pp. 67, 69]) is

$$S' = -aS + br(t)\frac{M - S}{M}.$$

This is a linear equation, which in the case where r is constant, also has constant coefficients. It is interesting to compare the solutions satisfying the initial condition $S(0) = S_0$, for different values of the parameters b, r (constant) and M.

2. Assume that the evolution of a population is governed by the logistic equation $p' = kp(1 - p/K)$. If that population is the target of harvesting (hunting, fishing or predators), then a diminishing of the growth rate occurs. Assume that at any time, a fraction rp of the population is removed this way. Then the dynamics of the population is described by the equation

$$p' = kp\left(1 - \frac{p}{K}\right) - rp,$$

where r is the per capita harvesting rate expressing the harvesting intensity ([15, p. 45]). Find the solution of the Cauchy problem with the initial condition $p(0) = p_0$ and analyze it for different values of parameter r. Is there an extinction risk for the population?

Follow the same program, also using computer visualization, assuming a constant rate of harvesting H, that is, for the equation ([15, p. 41]):

$$p' = kp\left(1 - \frac{p}{K}\right) - H.$$

Do the same for Malthus's equation and compare the solutions for various values of the parameters r and H.

3. In the mathematical models from above, the stimulating/inhibiting terms of the growth rate have an affine dependence on the dependent variable. However, many real situations ask for nonlinear corrections of the growth rate. Such an example is given by the equation

$$p' = kp\left(1 - \frac{p}{K}\right) - H(p),$$

where $H(p) = ap^2/(p^2 + b^2)$ ([15, pp. 38, 44]). This equation has served as a model for an ecological study of spruce budworm outbreaks in Canadian forests. The term $H(p)$ simulates the inhibiting effect due to some budworm-linked species of birds. Notice that $0 \leq H(p) \leq a$. Make some other comments on the bird-predation rate $H(p)$ and give an interpretation in ecological terms.

Seminar 3 Linear Differential Systems

3.1 Solved Exercises

Exercise 3.1. Consider the system

$$\begin{cases} x' = x + 2y \\ y' = 3x + 2y. \end{cases}$$

(a) Check that

$$u_1 = \begin{bmatrix} e^{-t} \\ -e^{-t} \end{bmatrix}, \quad u_2 = \begin{bmatrix} 2e^{4t} \\ 3e^{4t} \end{bmatrix}$$

are solutions of the system.
(b) Write down a fundamental matrix of the system.
(c) Find the general solution of the system.
(d) Find the solution that satisfies the conditions $x(0) = 1, y(0) = 9$.

Solution. (a) Check the result by direct substitution into the system.
(b) The solutions u_1, u_2 are linearly independent. Hence the matrix whose columns are u_1 and u_2 is a fundamental matrix of the system:

$$U(t) = \begin{bmatrix} e^{-t} & 2e^{4t} \\ -e^{-t} & 3e^{4t} \end{bmatrix}.$$

(c) The system is linear and homogeneous. Hence its general solution is $u = U(t)C$, where $C = [C_1, C_2]^T$ is arbitrary, or explicitly

$$\begin{cases} x = C_1 e^{-t} + 2C_2 e^{4t} \\ y = -C_1 e^{-t} + 3C_2 e^{4t}. \end{cases}$$

(d) Let $u_0 = [1, 9]^T$. Then the required conditions yield $U(0)C = u_0$, so that $C = U(0)^{-1} u_0$. Therefore, the solution of the Cauchy problem is $u = U(t)U(0)^{-1} u_0$.

Exercise 3.2. Consider the system $u' = Au + b(t)$, where

$$u = \begin{bmatrix} x \\ y \end{bmatrix}, \quad A = \begin{bmatrix} -3 & 8 \\ 1 & -1 \end{bmatrix}, \quad b(t) = \begin{bmatrix} -t \\ 1 \end{bmatrix}.$$

(a) Check that

$$u_1 = \begin{bmatrix} 2e^t \\ e^t \end{bmatrix}, \quad u_2 = \begin{bmatrix} -4e^{-5t} \\ e^{-5t} \end{bmatrix}$$

are solutions of the associated homogeneous system.
(b) Find a fundamental matrix $U(t)$ of the system.
(c) Write down the general solution of the system in terms of $U(t)$.

https://doi.org/10.1515/9783110447446-009

(d) Give the representation in terms of $U(t)$ of the solution of the Cauchy problem with the initial condition $u(0) = u_0$.

(e) Determine the fundamental matrix of the system $V(t)$ that satisfies the condition $V(0) = M$, where

$$M = \begin{bmatrix} -4 & 2 \\ 1 & 1 \end{bmatrix}.$$

Solution. (a) Check the result by direct substitution.

(b) The solutions u_1, u_2 are linearly independent. Hence a fundamental matrix is

$$U(t) = \begin{bmatrix} 2e^t & -4e^{-5t} \\ e^t & e^{-5t} \end{bmatrix}.$$

(c) The representation of the general solution is

$$u = U(t)C + \int_0^t U(t)U(s)^{-1} b(s)\, ds,$$

where $C = [C_1, C_2]^T$ is arbitrary.

(d) The solution of the Cauchy problem is

$$u = U(t)U(0)^{-1} u_0 + \int_0^t U(t)U(s)^{-1} b(s)\, ds.$$

(e) From (2.23), $V(t) = U(t)U(0)^{-1} V(0)$. Hence $V(t) = U(t)U(0)^{-1} M$. One has

$$U(0) = \begin{bmatrix} 2 & -4 \\ 1 & 1 \end{bmatrix}, \quad U(0)^{-1} = \frac{1}{6}\begin{bmatrix} 1 & 4 \\ -1 & 2 \end{bmatrix},$$

$$U(t)U(0)^{-1} = \frac{1}{6}\begin{bmatrix} 2e^t + 4e^{-5t} & 8e^t - 8e^{-5t} \\ e^t - e^{-5t} & 4e^t + 2e^{-5t} \end{bmatrix},$$

and after some calculation one gets

$$V(t) = \begin{bmatrix} -3e^{-5t} & 2e^t \\ e^{-5t} & e^t \end{bmatrix}.$$

Exercise 3.3. Solve the Cauchy problem $u' = Au + b$, $u(0) = u_0$, where

$$A = \begin{bmatrix} 3 & -1 \\ 1 & 1 \end{bmatrix}, \quad b = \begin{bmatrix} 1 \\ 2 \end{bmatrix}, \quad u_0 = \begin{bmatrix} 1 \\ 2 \end{bmatrix},$$

if it is known that

$$U(t) = e^{2t}\begin{bmatrix} 1+t & -t \\ t & 1-t \end{bmatrix}$$

is a fundamental matrix of the system.

Solution. Use the formula

$$u = U(t)\left(U(0)^{-1}u_0 + \int_0^t U(s)^{-1}b(s)\,ds\right).$$

One has $U(0) = I = U(0)^{-1}$. Also, $\det U(t) = e^{4t}$ and

$$U(t)^{-1} = e^{-2t}\begin{bmatrix} 1-t & t \\ -t & 1+t \end{bmatrix}.$$

Then

$$U(t)^{-1}b(t) = e^{-2t}\begin{bmatrix} 1+t \\ 2+t \end{bmatrix},$$

$$\int_0^t U(s)^{-1}b(s)\,ds = \begin{bmatrix} -\frac{2t+3}{4}e^{-2t} + \frac{3}{4} \\ -\frac{2t+5}{4}e^{-2t} + \frac{5}{4} \end{bmatrix}.$$

Furthermore, by direct calculation,

$$U(0)^{-1}u_0 + \int_0^t U(s)^{-1}b(s)\,ds = \begin{bmatrix} -\frac{2t+3}{4}e^{-2t} + \frac{7}{4} \\ -\frac{2t+5}{4}e^{-2t} + \frac{13}{4} \end{bmatrix}$$

and finally,

$$u = U(t)\begin{bmatrix} -\frac{2t+3}{4}e^{-2t} + \frac{7}{4} \\ -\frac{2t+5}{4}e^{-2t} + \frac{11}{4} \end{bmatrix} = e^{2t}\begin{bmatrix} \frac{7}{4}te^{-2t} - \frac{3}{4}e^{-2t} - \frac{11}{4}t + \frac{7}{4} \\ -\frac{5}{4}e^{-2t} - \frac{3}{2}t + \frac{13}{4} \end{bmatrix}.$$

Exercise 3.4. Determine a particular solution of the system

$$\begin{cases} x' = x + 3y + 1 \\ y' = -x + t. \end{cases}$$

Solution. The nonhomogeneous terms 1 and t are polynomials of maximum degree 1. Thus we guess a particular solution given by first-degree polynomials $u_p = [at+b, ct+d]^T$. Substituting into the system and identifying coefficients yields $a = 1$, $b = 1/3$, $c = -1/3$, $d = -1/9$. Therefore, a particular solution of the system is

$$u_p = \left[t + \frac{1}{3},\ -\frac{1}{3}t - \frac{1}{9}\right]^T.$$

The method that we used is called the *method of undetermined coefficients*.

Exercise 3.5. Using the method of characteristic equation solve the system $u' = Au$ in each of the following cases:

(a) $A = \begin{bmatrix} 5 & -3 \\ 8 & -6 \end{bmatrix}$; (b) $A = \begin{bmatrix} -3 & 4 \\ -1 & 1 \end{bmatrix}$; (c) $A = \begin{bmatrix} 1 & -4 \\ 2 & 5 \end{bmatrix}$.

Solution. (a) Find solutions of the form $u = e^{rt}v$, where $v = [v_1, v_2]^T$. The characteristic equation is $r^2 - (\operatorname{tr} A)r + \det A = 0$. In this case $r^2 + r - 6 = 0$. The roots are $r_1 = -3$, $r_2 = 2$. Now we look for an eigenvector v by solving the homogeneous algebraic system $(A - rI)v = 0$. To this aim, it suffices to consider only its first equation. For $r = -3$, one has $8v_1 - 3v_2 = 0$, and choosing $v_1 = 3$ we obtain $v_2 = 8$. Hence a solution of the differential system is $u_1 = [3e^{-3t}, 8e^{-3t}]^T$. For $r = 2$, one has $3v_1 - 3v_2 = 0$ and we may choose the solution $v_1 = v_2 = 1$, which gives a second solution of the differential system, namely $u_2 = [e^{2t}, e^{2t}]^T$. Then, the general solution of the differential system is $u = C_1 u_1 + C_2 u_2$, where C_1, C_2 are arbitrary real numbers, or explicitly

$$\begin{cases} x = 3C_1 e^{-3t} + C_2 e^{2t} \\ y = 8C_1 e^{-3t} + C_2 e^{2t} \ . \end{cases}$$

(b) The characteristic equation is $r^2 + 2r + 1 = 0$, with roots $r_1 = r_2 = -1$. Two solutions are

$$u_1 = e^{rt}v, \quad u_2 = e^{rt}(tv + w) \ ,$$

where $(A - rI)v = 0$ and $(A - rI)w = v$. The first equation of the algebraic system in unknown v is $-2v_1 + 4v_2 = 0$, and one of its solutions is $v_1 = 2$, $v_2 = 1$. Next the first equation of the second algebraic system (in w) is $-2w_1 + 4w_2 = 2$, and one of its solutions is $w_1 = w_2 = 1$. Thus we have obtained the solutions

$$u_1 = \begin{bmatrix} 2e^{-t} \\ e^{-t} \end{bmatrix}, \quad u_2 = \begin{bmatrix} (2t+1)e^{-t} \\ (t+1)e^{-t} \end{bmatrix}.$$

The general solution of the system is

$$\begin{cases} x = [2C_1 + C_2(2t+1)] \, e^{-t} \\ y = [C_1 + C_2(t+1)] \, e^{-t} \ . \end{cases}$$

(c) The roots of the characteristic equation $r^2 - 6r + 13 = 0$ are $3 \pm 2i$. We solve in complex numbers the algebraic system $(A - rI)v = 0$. One has $(-2 - 2i)v_1 - 4v_2 = 0$, which with the choice $v_1 = 2$ gives $v_2 = -1 - i$. Thus we have obtained a complex-valued solution of the differential system

$$u = e^{(3+2i)t} \begin{bmatrix} 2 \\ -1-i \end{bmatrix} = e^{3t}(\cos 2t + i \sin 2t) \left(\begin{bmatrix} 2 \\ -1 \end{bmatrix} + i \begin{bmatrix} 0 \\ -1 \end{bmatrix} \right).$$

Two real-valued solutions are then given by the real and imaginary parts,

$$u_1 = e^{3t} \left(\cos 2t \begin{bmatrix} 2 \\ -1 \end{bmatrix} - \sin 2t \begin{bmatrix} 0 \\ -1 \end{bmatrix} \right) = \begin{bmatrix} 2e^{3t} \cos 2t \\ e^{3t}(-\cos 2t + \sin 2t) \end{bmatrix},$$

$$u_2 = e^{3t} \left(\cos 2t \begin{bmatrix} 0 \\ -1 \end{bmatrix} + \sin 2t \begin{bmatrix} 2 \\ -1 \end{bmatrix} \right) = \begin{bmatrix} 2e^{3t} \sin 2t \\ -e^{3t}(\cos 2t + \sin 2t) \end{bmatrix}.$$

Exercise 3.6. Consider the system

$$u' = \begin{bmatrix} -1 & 2 \\ 2 & -4 \end{bmatrix} u.$$

(a) Find the general solution.
(b) Show that the orbits of the system are semilines.
(c) Sketch the phase portrait of the system, being sure to indicate the directions of the orbits.

Solution. (a) The roots of the characteristic equation $r^2 + 5r = 0$ are $r_1 = -5, r_2 = 0$. The first equation of the algebraic system $(A - rI)v = 0$, for $r = -5$, is $4v_1 + 2v_2 = 0$. Thus we may choose $v_1 = 1, v_2 = -2$. For $r = 0$, one has the equation $-v_1 + 2v_2 = 0$, with one solution $v_1 = 2, v_2 = 1$. Hence two solutions of the differential system are

$$u_1 = e^{-5t}\begin{bmatrix} 1 \\ -2 \end{bmatrix}, \quad u_2 = \begin{bmatrix} 2 \\ 1 \end{bmatrix}.$$

Therefore, the general solution is

$$\begin{cases} x = C_1 e^{-5t} + 2C_2 \\ y = -2C_1 e^{-5t} + C_2. \end{cases} \quad (3.1)$$

(b) The equalities (3.1) are the parametric equations of the orbits. Eliminating t yields the Cartesian equations of the orbits,

$$y + 2x = 5C_2,$$

which represent a pencil of parallel lines of slope -2. Since $x \to 2C_2$ and $y \to C_2$ as $t \to +\infty$, the orbits are semilines having origin in $(2C_2, C_2)$, and the direction is towards the s emiline origin.

3.2 Proposed Exercises

Exercise 3.7. Consider the system

$$\begin{cases} x' = -2x + y \\ y' = 4x + y. \end{cases}$$

(a) Verify that

$$u_1 = \begin{bmatrix} e^{2t} \\ 4e^{2t} \end{bmatrix}, \quad u_2 = \begin{bmatrix} e^{-3t} \\ -e^{-3t} \end{bmatrix}$$

are solutions of the system.
(b) Write down a fundamental matrix $U(t)$ of the system.

(c) Find the general solution.
(d) Solve the Cauchy problem with the initial conditions $x(1) = -1$, $y(1) = 1$.
(e) Find the fundamental matrix e^{tA}, where A is the matrix of the system coefficients.

Exercise 3.8. Use the method of undetermined coefficients to determine a particular solution in the form of $b(t)$, for the system $u' = Au + b(t)$, where

$$A = \begin{bmatrix} 4 & -3 \\ 2 & 1 \end{bmatrix},$$

in each one of the following cases:

(a) $b(t) = \begin{bmatrix} e^t \\ 1 \end{bmatrix}$; (b) $b(t) = \begin{bmatrix} -2 \\ t \end{bmatrix}$; (c) $b(t) = \begin{bmatrix} 0 \\ t^2 \end{bmatrix}$.

Exercise 3.9. Find a particular solution of the system

$$\begin{cases} x' = x - y + 1 + e^t \\ y' = x + y - t. \end{cases}$$

Exercise 3.10. Solve the system $u' = Au$ in each one of the following cases:

(a) $A = \begin{bmatrix} 2 & 0 \\ 1 & 2 \end{bmatrix}$; (b) $A = \begin{bmatrix} 0 & 4 \\ 5 & 1 \end{bmatrix}$; (c) $A = \begin{bmatrix} 0 & -\frac{1}{2} \\ 5 & 1 \end{bmatrix}$.

Exercise 3.11. Find the general solution of the system $u' = Au$, the Cartesian equations of the orbits, and sketch the phase portrait, in each one of the following cases:

(a) $A = \begin{bmatrix} 3 & 0 \\ 0 & 3 \end{bmatrix}$; (b) $A = \begin{bmatrix} -1 & 0 \\ 0 & 1 \end{bmatrix}$;

(c) $A = \begin{bmatrix} 0 & 1 \\ 1 & 0 \end{bmatrix}$; (d) $A = \begin{bmatrix} 0 & -2 \\ 2 & 0 \end{bmatrix}$.

Exercise 3.12. (a) Find the values of the parameter a for which the system

$$\begin{cases} x' = ax - 5y \\ y' = x - 2y \end{cases}$$

has periodic solutions (closed orbits), and then find the solutions.
(b) Find the values of a for which the orbits belong to a pencil of parallel lines.

Exercise 3.13. Characterize those homogeneous linear planar systems with constant coefficients whose orbits belong to a pencil of lines through the origin.

Exercise 3.14. Study all possible behaviors of the orbits of the system

$$\begin{cases} x' = ax - y \\ y' = x + ay. \end{cases}$$

3.3 Hints and Solutions

Exercise 3.7
(c) $x = C_1 e^{2t} + C_2 e^{-3t}$, $\quad y = 4C_1 e^{2t} - C_2 e^{-3t}$;
(d) $u = U(t) U(1)^{-1} u_0$.

Exercise 3.8
(a) $x = -e^t/2$, $\quad y = -e^t/6$;
(b) $x = -3t/10 + 1/20$, $\quad y = -2t/5 - 1/2$;
(c) $x = -3t^2/10 - 3t/10 - 9/100$, $\quad y = -2t^2/5 - t/5 - 1/50$.

Exercise 3.9
$u = u_1 + u_2$, where u_1, u_2 are particular solutions of the systems

$$x' = x - y + 1, \quad y' = x + y - t; \qquad x' = x - y + e^t, \quad y' = x + y.$$

One obtains $u_1 = [t/2, (1+t)/2]^T$ and $u_2 = [0, e^t]^T$. Hence

$$x = \frac{1}{2} t, \quad y = \frac{1}{2}(1 + t) + e^t.$$

Exercise 3.10
(a) $x = C_2 e^{2t}$, $\quad y = (C_1 + C_2 t) e^{2t}$;
(b) $x = 4C_1 e^{5t} + C_2 e^{-4t}$, $\quad y = 5C_1 e^{5t} - C_2 e^{-4t}$;
(c) $x = e^{\frac{t}{2}} \left(C_1 \sin \frac{3t}{2} + C_2 \cos \frac{3t}{2} \right)$,
$y = -e^{\frac{t}{2}} \left[C_1 \left(\sin \frac{3t}{2} + 3 \cos \frac{3t}{2} \right) + C_2 \left(\cos \frac{3t}{2} - 3 \sin \frac{3t}{2} \right) \right]$.

Exercise 3.11
(a) Semilines coming out from the origin.
(b) $xy = c$ (equilateral hyperbolas).
(c) $x^2 - y^2 = c$ (hyperbolas).
(d) $x^2 + y^2 = c$ (concentric circles).

Exercise 3.12
(a) The solutions are periodic if and only if the roots of the characteristic equations have the form $\pm i\beta$, with $\beta \neq 0$, that is, $\operatorname{tr} A = 0$ and $\det A > 0$. Then $a = 2$, and the solutions are:

$$x = 5 (C_1 \cos t + C_2 \sin t), \quad y = C_1 (2 \cos t + \sin t) + C_2 (2 \sin t - \cos t).$$

(b) The orbits belong to a pencil of parallel lines, that is, their Cartesian equations are of the form $\alpha x + \beta y + \gamma = 0$, where α, β are constants that are not both zero, if and only if $\alpha x' + \beta y' = 0$. Hence $\alpha(ax - 5y) + \beta(x - 2y) = 0$. It follows that $a\alpha + \beta = 0$ and $-5\alpha - 2\beta = 0$. This system has nonzero solutions only if its determinant is equal to zero. Then $a = 5/2$.

Exercise 3.13
If the orbits have Cartesian equations of the form $y = Cx$, then $(y/x)' = 0$, that is $y'x - yx' = 0$. This happens only for differential systems of the type $x' = ax$, $y' = ay$.

Exercise 3.14
The roots of the characteristic equation are $a \pm i$. The orbits are spirals approaching the origin if $a < 0$; spirals receding from the origin if $a > 0$; closed curves if $a = 0$.

3.4 Project: Mathematical Models Represented by Linear Differential Systems

Suggested structure. 1. Glucose-insulin dynamics; 2. Drug concentration in blood and body tissues; 3. Chemical reaction dynamics; 4. Contamination of reservoirs in a network; 5. Metastasis of malignant tumors.

Basic bibliography. V. Barbu [2], D. Kaplan and L. Glass [13], J. D. Logan [15].

Hints. 1. [13, p. 261], [15, p. 197] Let $x(t)$ and $y(t)$ be the displacements from the basal values of the glucose and insulin respectively at time t (hours). A simple model of the glucose-insulin interaction is given by the linear differential system

$$\begin{cases} x' = -ax - by \\ y' = cx - dy, \end{cases}$$

where ax is the rate glucose is absorbed in the liver, by is the rate glucose is used in the muscle, cx is the rate insulin is produced by the pancreas, and dy is rate insulin is degraded by the liver.

Plot the phase portrait of the system for different sets of values for the parameters a, b, c, d around the values $3, 4.3, 0.2, 0.8$ respectively.

2. [13, p. 263], [15, p. 203] In pharmaceutical research it is useful to have a model that allows one to find the proportion of an administered drug that effectively reaches the target organ or body tissues. Let $x(t)$ and $y(t)$ denote the amounts of the drug in blood and in tissues respectively. Then the system that describes, also for other similar processes, the dynamics of the flow of a substance from one compartment into another (here from blood into tissues) is the bicompartmental model

$$\begin{cases} x' = -ax - cx + by \\ y' = ax - by. \end{cases}$$

The term ax is the amount of the drug passing from blood into tissues, by is the rate the drug passes from tissues into blood, and cx is the rate the drug is degraded by the liver.

Plot the phase portrait of the system for different sets of values for the parameters a, b, c around the values $0.5, 0.5, 2$ respectively.

3.4 Project: Mathematical Models Represented by Linear Differential Systems

3. See [13, p. 265].

4. [15, p. 202] Let V_1 and V_2 be the volumes of two connected water reservoirs, initially not contaminated. A toxic chemical flows into reservoir I at flow rate $q + r$ (liters/minute) and concentration c (grams/liter). From reservoir I, a portion of the contaminated water flows into reservoir II at flow rate q, and another portion is eliminated by a draining pipe at flow rate r. In order to maintain a constant volume of water in reservoir II, water is eliminated at a flow rate equal to the entrance one, namely q. If $x(t)$ and $y(t)$ denote the concentrations of the chemical in reservoir I and reservoir II respectively, then the dynamics of contamination of the two reservoirs is described by the linear differential system

$$\begin{cases} V_1 x' = (q+r)c - qx - rx \\ V_2 y' = qx - qy \,. \end{cases}$$

Solve the system under initial conditions $x(0) = x_0$ and $y(0) = y_0$; find the equilibrium concentrations in the two reservoirs; for different sets of values for parameters, plot the corresponding time series representations of the solutions.

5. See [13, pp. 221–224].

Seminar 4 Second-Order Differential Equations

4.1 Solved Exercises

Exercise 4.1. Solve the following equations by reduction of order:

(a) $x'' = \dfrac{3}{t} x'$; (b) $x'' = x^2 x'$; (c) $x'' = -8x$.

Solution. (a) Let $y = x'$. Then $y' = 3y/t$, and by separating variables and integrating we obtain $y = C_1 t^3$. This gives $x = C_1 t^4/4 + C_2$, where $C_1, C_2 \in \mathbb{R}$.

(b) For $y = x'$ we have
$$x'' = \frac{dy}{dt} = \frac{dy}{dx}\frac{dx}{dt} = y\frac{dy}{dx}.$$
Substituting into the equation yields $y\,dy/dx = x^2 y$. Hence $y = 0$ or $dy/dx = x^2$, that is, $y = (x^3 + C_1)/3$. Therefore the equation is equivalent to the first-order equations $x' = 0$ and $x' = (x^3 + C_1)/3$.

(c) Multiply by x' to obtain the energy conservation theorem $x'^2 + 8x^2 = 2E$.

Exercise 4.2. Determine the total energy corresponding to the equation of motion $x'' = -x$, if the initial position and velocity are $x(0) = 0$ and $x'(0) = 1$ respectively. Then find the law of motion.

Solution. One has $x'^2 + x^2 = 2E$. For $t = 0$, this gives $E = 1/2$. The law of motion is obtained by solving the equation $x'^2 + x^2 = 1$. One has $x' = \pm\sqrt{1-x^2}$. Since $x'(0) > 0$, only the equation with $+$ is convenient. Furthermore, separating variables yields $\arcsin x = t + C$. Using $x(0) = 0$ we deduce $C = 0$ and finally $x = \sin t$.

Exercise 4.3. Solve the following equations with constant coefficients:
(a) $x'' + 5x' + 6x = 1 + 12t$;
(b) $x'' - 6x' + 9x = 7e^t$;
(c) $x'' - 10x' + 29x = 39 - 29t$.

Solution. (a) The characteristic equation $r^2 + 5r + 6 = 0$ has the roots $r_1 = -3$ and $r_2 = -2$. Hence the general solution of the homogeneous equation is $x_h = C_1 e^{-3t} + C_2 e^{-2t}$. A particular solution of the nonhomogeneous equation is sought in the form $x_p = at + b$. It follows that $x_p = 2t - 3/2$. Consequently $x = C_1 e^{-3t} + C_2 e^{-2t} + 2t - 3/2$.

(b) The characteristic equation $r^2 - 6r + 9 = 0$ has the double root $r = 3$. Hence $x_h = (C_1 + C_2 t)e^{3t}$. Looking for a particular solution $x_p = ae^t$, we obtain $x_p = 7e^t/4$. Therefore $x = (C_1 + C_2 t)e^{3t} + 7e^t/4$.

(c) The roots of the characteristic equation $r^2 - 10r + 29 = 0$ are $5 \pm 2i$. Hence $x_h = e^{5t}(C_1 \cos 2t + C_2 \sin 2t)$. A particular solution having the form of the right-hand side of the equation is $x_p = -t + 1$. Therefore $x = e^{5t}(C_1 \cos 2t + C_2 \sin 2t) - t + 1$.

Exercise 4.4. For the following equations:

$$\text{(i) } x'' - 4x = f(t); \qquad \text{(ii) } x'' + 4x = f(t),$$

give the form of a particular solution in each of the cases: (a) $f = te^t$; (b) $f = (t+1)e^{2t}$; (c) $f = (1-t)\sin 2t$.

Solution. (i) (a) $x_p = (at+b)e^t$;
(b) $x_p = t(at+b)e^{2t}$; the factor t in front is due to the fact that $r = 2$ is a simple root of the characteristic equation;
(c) $x_p = (at+b)\cos 2t + (ct+d)\sin 2t$.
(ii) (a) $x_p = (at+b)e^t$;
(b) $x_p = (at+b)e^{2t}$;
(c) $x_p = t[(at+b)\cos 2t + (ct+d)\sin 2t]$ since $r = 2i$ is a simple root of the characteristic equation.

Exercise 4.5. Solve the equation

$$Lx := x^{(VI)} - 3x^{(V)} + 6x''' - 3x'' - 3x' + 2x = 0$$

and indicate the form of a particular solution of the equation $Lx = f$, in each of the following cases: (a) $f = (1-t)e^{5t}$; (b) $f = (1-t)e^t$; (c) $f = (1-t)e^{-t}$; (d) $f = (1-t)e^{2t}$.

Solution. The characteristic equation $r^6 - 3r^5 + 6r^3 - 3r^2 - 3r + 2 = 0$ has the roots $r_1 = r_2 = r_3 = 1$, $r_4 = r_5 = -1$ and $r_6 = 2$. Hence the general solution of the equation $Lx = 0$ is

$$x = (C_1 + C_2 t + C_3 t^2)e^t + (C_4 + C_5 t)e^{-t} + C_6 e^{2t},$$

where $C_1, C_2, \ldots, C_6$ are arbitrary constants.
A particular solution of the equation $Lx = f$ is sought in the form: (a) $x_p = (at+b)e^{5t}$; (b) $x_p = t^3(at+b)e^t$ since the root $r = 1$ is triple; (c) $x_p = t^2(at+b)e^{-t}$ since the root $r = -1$ is double; (d) $x_p = t(at+b)e^{2t}$ since the root $r = 2$ is simple.

Exercise 4.6. Solve the Euler equations:

$$\text{(a) } t^2 x'' - tx' - 3x = 0; \qquad \text{(b) } t^2 x'' - tx' - 3x = 3\ln t - 2 \quad (t > 0).$$

Solution. (a) The equation being homogeneous, we may look for solutions of the form t^α. Substituting yields $\alpha(\alpha - 1) - \alpha - 3 = 0$, or $\alpha^2 - 2\alpha - 3 = 0$. Then $\alpha = -1$ and $\alpha = 3$, and the general solution of the equation is $x = C_1 t^{-1} + C_2 t^3$.

(b) One method is to look for a particular solution of the nonhomogeneous equation under the form of the right-hand side (exercise). Another method, which is applicable to the general case, consists of making the change of variable $s = \ln t$. Then

$$x' = \frac{dx}{dt} = \frac{dx}{ds}\frac{ds}{dt} = \frac{1}{t}\frac{dx}{ds},$$

$$x'' = \frac{d}{dt}\left(\frac{1}{t}\frac{dx}{ds}\right) = -\frac{1}{t^2}\frac{dx}{ds} + \frac{1}{t}\frac{d}{dt}\left(\frac{dx}{ds}\right)$$
$$= -\frac{1}{t^2}\frac{dx}{ds} + \frac{1}{t}\frac{d^2x\,ds}{ds^2\,dt} = -\frac{1}{t^2}\frac{dx}{ds} + \frac{1}{t^2}\frac{d^2x}{ds^2}.$$

Thus the equation becomes a constant coefficient equation

$$\frac{d^2x}{ds^2} - 2\frac{dx}{ds} - 3x = 3s - 2.$$

Solving it we find $x = C_1 e^{-s} + C_2 e^{3s} - s + 4/3$. Finally, coming back to the variable t we obtain $x = C_1 t^{-1} + C_2 t^3 - \ln t + 4/3$.

Exercise 4.7. Solve the following system using the method of elimination:

$$\begin{cases} x' = 2x - y + t \\ y' = -6x + y - 1. \end{cases}$$

Solution. Differentiating into the first equation yields $x'' = 2x' - y' + 1$. Replace y' using the second equation and find $x'' = 2x' + 6x - y + 2$. Now using the first equation of the system, replace y by $-x' + 2x + t$ and find $x'' = 2x' + 6x + x' - 2x - t + 2$, or

$$x'' - 3x' - 4x = -t + 2.$$

Solving it gives $x = C_1 e^{-t} + C_2 e^{4t} + t/4 - 11/16$. Next $y = 3C_1 e^{-t} - 2C_2 e^{4t} + 3t/2 - 13/8$.

4.2 Proposed Exercises

Exercise 4.8. Reduce the order of the following equations:
(a) $x'' = t^5 + x'$;
(b) $x'' = x^2 + x'^2$;
(c) $x'' = 2x - x^2$.

Exercise 4.9. Find the value of total energy corresponding to the equation of motion $x'' = 2\sqrt{x}$, if the initial position and velocity are $x(0) = 1$ and $x'(0) = 2$.

Exercise 4.10. Solve the following linear equations with constant coefficients:
(a) $x'' - x = t + 3e^{2t}$;
(b) $4x'' - 4x' + x = 1 - e^t$;
(c) $x'' + x = \cos 3t - 1 + t^2$;
(d) $x'' - 4x' + 5x = 5e^{3t} + 16\sin t$.

Exercise 4.11. For the following equations:
(i) $4x'' - x = f(t)$;
(ii) $4x'' + x = f(t)$,

indicate the form of a particular solution in each of the cases:

$$(a)\, f = te^t; \quad (b)\, f = (t+1)e^{\frac{t}{2}}; \quad (c)\, f = (1-t)\sin\frac{t}{2}.$$

Exercise 4.12. Solve the equations:
(a) $x''' - 2x'' - 15x' + 36x = 0$;
(b) $x^{(V)} - x''' = 0$;
(c) $x^{(IV)} + 2x'' + x = 0$.

Exercise 4.13. Find the solution of the Cauchy problem

$$\begin{cases} 2x''' + x'' - 2x' - x = t \\ x(0) = 6, \quad x'(0) = -5, \quad x''(0) = 1. \end{cases}$$

Exercise 4.14. Solve the initial value problems:
(a) $t^2 x'' - 3tx' + 8x = 0 \; (t > 0), \quad x(1) = 0, \quad x'(1) = -2$;
(b) $x^2 y'' + xy' - 2y = 0 \; (x > 0), \quad y(1) = 0, \quad y'(1) = -1$ (here $y = y(x)$).

4.3 Solutions

Exercise 4.8
(a) $y' = t^5 + y$;
(b) $dy/dx = x^2 y^{-1} + y$;
(c) $x'^2/2 - x^2 + x^3/3 = E$.

Exercise 4.9
$E = 2/3$.

Exercise 4.10
(a) $x = C_1 e^t + C_2 e^{-t} - t + e^{2t}$;
(b) $x = (C_1 + C_2 t)e^{\frac{t}{2}} + 1 - e^t$;
(c) $x = C_1 \cos t + C_2 \sin t + t^2 - 3$;
(d) $x = (C_1 \cos t + C_2 \sin t)e^{2t} + 5e^{3t}/2 + 2\cos t + 2\sin t$.

Exercise 4.11
(i) (a) $x_p = (at+b)e^t$;
    (b) $x_p = (at^2 + bt)e^{t/2}$;
    (c) $x_p = (at+b)\cos(t/2) + (ct+d)\sin(t/2)$;
(ii) (a) $x_p = (at+b)e^t$;
    (b) $x_p = (at+b)e^{t/2}$;
    (c) $x_p = (at^2 + bt)\cos(t/2) + (ct^2 + dt)\sin(t/2)$.

Exercise 4.12
(a) $x = (C_1 + C_2 t)e^{3t} + C_3 e^{-4t}$
(b) $x = C_1 + C_2 t + C_3 t^2 + C_4 e^t + C_5 e^{-t}$;
(c) $x = (C_1 + C_2 t)\cos t + (C_3 + C_4 t)\sin t$.

Exercise 4.13
$x = e^{-t} + 4e^{-t/2} - e^t - t + 2$.

Exercise 4.14
(a) $x = -t^2 \sin(2 \ln t)$;
(b) $y = \sqrt{2}\left(x^{-\sqrt{2}} - x^{\sqrt{2}}\right)/4$.

4.4 Project: Boundary Value Problems for Second-Order Differential Equations

Suggested structure. 1. Boundary value problems as models of physical processes; 2. Green's function. Examples.

Basic bibliography. R. P. Agarwal and D. O'Regan [1], J. D. Logan [15], L. C. Piccinini, G. Stampacchia and G. Vidossich [19].

Hints. 1. Stationary heat conduction in a metallic bar [15, pp. 117–120]. Consider a thin cylindrical uniform metallic bar of length l and of cross-sectional area A. The axis of the cylinder is identified with the segment $[0, l]$ of the real line Ox. Assume that the bar is thermally insulated on its lateral side, and that its ends are maintained at constant temperatures, u_0 at the left end $x = 0$, and u_l at the right end $x = l$. Also assume the existence of heat sources inside the bar. The cylinder being sufficiently thin, we may accept that the heat transfer occurs along the axis Ox and that the temperature in the bar is a function $u(x)$ depending only on the abscissa x. We are interested in the stationary heat distribution inside the bar, that is, after thermal equilibrium has been attained. In order to obtain the mathematical model of this process, let us denote by $\phi(x)$ the heat flux density (heat quantity through the cross-section of abscissa x, per unit area and unit time), and by $f(x)$ the source density at x (heat quantity internally produced by the sources per unit volume and unit time). Then, for a small section $[x, x + dx]$ of the bar, we may write down the law of conservation of energy: the heat quantity that flows in, plus the heat generated by sources, equals the heat quantity that flows out that segment, that is

$$A\phi(x) + f(x)A\,dx = A\phi(x + dx).$$

Then

$$\frac{\phi(x + dx) - \phi(x)}{dx} = f(x),$$

4.4 Project: Boundary Value Problems for Second-Order Differential Equations

where letting $dx \to 0$ gives

$$\phi'(x) = f(x).$$

If we denote by $u(x)$ the temperature at section x and we agree that the flux density $\phi(x)$ is proportional to $u'(x)$, more exactly

$$\phi(x) = -k(x)u'(x) \quad \text{(Fourier's heat conduction law),}$$

then we obtain the stationary heat conduction equation

$$-\bigl(k(x)u'(x)\bigr)' = f(x),$$

explicitly

$$-k(x)u'' - k'(x)u' = f(x).$$

The coefficient $k(x)$ is called the thermal conductivity. For a uniform bar, the thermal conductivity is a constant k, and the equation reduces to

$$-ku'' = f(x).$$

If we add to the heat equation the conditions at the ends, called *boundary conditions*, then we obtain the *boundary value problem*

$$\begin{cases} -k(x)u'' - k'(x)u' = f(x), & 0 \le x \le l \\ u(0) = u_0, \quad u(l) = u_l. \end{cases}$$

Other boundary conditions are possible, for example: $u'(0) = \alpha$ and $u'(l) = \beta$ (giving the flux at the ends of the bar).

2. Determine Green's function (Section I.3.6) for different boundary conditions.

Seminar 5 Gronwall's Inequality

5.1 Solved Exercises

Exercise 5.1. Determine an upper bound of the function $y \in C[0, 1]$ that satisfies the inequality:
(a) $y(x) \leq 2 + 3 \int_0^x y(s)\, ds$ for all $x \in [0, 1]$;
(b) $y(x) \leq 2 + 3 \int_x^1 y(s)\, ds$ for all $x \in [0, 1]$;
(c) $y(x) \leq 3 \int_0^x y(s)\, ds$ for all $x \in [0, 1]$.

Solution. (a) Theorem 4.1 yields $y(x) \leq 2e^{3x}$ for every $x \in [0, 1]$.

(b) Theorem 4.2 implies that $y(x) \leq 2e^{3(1-x)}$ for every $x \in [0, 1]$.

(c) Theorem 4.1 guarantees $y(x) \leq 0$ for every $x \in [0, 1]$. It is useful to give a direct proof of this result. To this aim, let $z(x) = \int_0^x y(s)\, ds$. We have $z'(x) \leq 3z(x)$, which after multiplying by e^{-3x} becomes $z'(x)e^{-3x} - 3z(x)e^{-3x} \leq 0$, that is $(z(x)e^{-3x})' \leq 0$. Hence the function $z(x)e^{-3x}$ is nonincreasing on $[0, 1]$. Since $z(0) = 0$, we then have $z(x)e^{-3x} \leq 0$, hence $z(x) \leq 0$ for all $x \in [0, 1]$. Now the inequality from the hypothesis, $y(x) \leq 3z(x)$, implies that $y(x) \leq 0$ for all $x \in [0, 1]$.

Exercise 5.2. Let $x, y \in C^1[0, T]$ be two solutions of the equation

$$x' = t^2 + 2 \sin 3x\,.$$

Estimate the difference $x(t) - y(t)$ in terms of the initial values $x(0) = x_0$ and $y(0) = y_0$.

Solution. For every $t \in [0, T]$, one has

$$x(t) = x_0 + \int_0^t \left(s^2 + 2 \sin 3x(s)\right) ds,$$

$$y(t) = y_0 + \int_0^t \left(s^2 + 2 \sin 3y(s)\right) ds.$$

It follows that

$$|x(t) - y(t)| \leq |x_0 - y_0| + 2 \int_0^t |\sin 3x(s) - \sin 3y(s)|\, ds.$$

Using the mean value theorem, for some $c = c(s)$ between $3x(s)$ and $3y(s)$, we have

$$|\sin 3x(s) - \sin 3y(s)| = 3\,|x(s) - y(s)|\,|\cos c| \leq 3\,|x(s) - y(s)|\,.$$

Consequently,

$$|x(t) - y(t)| \leq |x_0 - y_0| + 6 \int_0^t |x(s) - y(s)|\, ds.$$

Now Gronwall's inequality gives
$$|x(t) - y(t)| \leq |x_0 - y_0| e^{6t}.$$
Finally, for every $t \in [0, T]$, we have
$$|x(t) - y(t)| \leq |x_0 - y_0| e^{6T}.$$

Exercise 5.3. Let $f, g \colon [0, T] \times \mathbb{R} \to \mathbb{R}$ be continuous. Assume that there exists a constant $L > 0$ such that $|g(t, u) - g(t, v)| \leq L|u - v|$ for all $t \in [0, T]$ and $u, v \in \mathbb{R}$. Prove that if $x, y \in C^1[0, T]$ solves the equations $x' = f(t, x)$ and $y' = g(t, y)$ respectively, then
$$|x(t) - y(t)| \leq (|x_0 - y_0| + \eta T) e^{Lt}$$
for every $t \in [0, T]$, where $x_0 = x(0)$, $y_0 = y(0)$ and
$$\eta = \max_{t \in [0,T]} |f(t, x(t)) - g(t, x(t))|.$$

Solution. From
$$x(t) = x_0 + \int_0^t f(s, x(s))\, ds, \quad y(t) = y_0 + \int_0^t g(s, y(s))\, ds,$$
we infer
$$|x(t) - y(t)|$$
$$\leq |x_0 - y_0| + \int_0^t |f(s, x(s)) - g(s, y(s))|\, ds$$
$$\leq |x_0 - y_0| + \int_0^t (|f(s, x(s)) - g(s, x(s))| + |g(s, x(s)) - g(s, y(s))|)\, ds$$
$$\leq |x_0 - y_0| + \eta T + L \int_0^t |x(s) - y(s)|\, ds.$$
Now the result follows from Gronwall's inequality.

Exercise 5.4. Assume that $f, g \colon [0, T] \times \mathbb{R}^2 \to \mathbb{R}$ are continuous functions and there exist constants $a, b, c, d \in \mathbb{R}_+$ such that
$$|f(t, x, y) - f(t, \bar{x}, \bar{y})| \leq a|x - \bar{x}| + b|y - \bar{y}|,$$
$$|g(t, x, y) - g(t, \bar{x}, \bar{y})| \leq c|x - \bar{x}| + d|y - \bar{y}|,$$
for all $t \in [0, T]$ and $x, \bar{x}, y, \bar{y} \in \mathbb{R}$. Let $\|\cdot\|$ be the Euclidean norm in $\mathbb{R}^2$ and let $u = (x, y)$ and $v = (\bar{x}, \bar{y})$ be two solutions on $[0, T]$ of the system
$$\begin{cases} x' = f(t, x, y) \\ y' = g(t, x, y). \end{cases}$$
Give an estimation for $\|u(t) - v(t)\|$ in terms of the initial values $u(0) = u_0$ and $v(0) = v_0$.

Solution. We have

$$|x(t) - \bar{x}(t)| \leq |x_0 - \bar{x}_0| + \int_0^t |f(s, x(s), y(s)) - f(s, \bar{x}(s), \bar{y}(s))|\, ds$$

$$\leq |x_0 - \bar{x}_0| + \int_0^t (a|x(s) - \bar{x}(s)| + b|y(s) - \bar{y}(s)|)\, ds.$$

Furthermore

$$|x(t) - \bar{x}(t)| \leq |x_0 - \bar{x}_0| + \int_0^t (a^2 + b^2)^{\frac{1}{2}} \left(|x(s) - \bar{x}(s)|^2 + |y(s) - \bar{y}(s)|^2\right)^{\frac{1}{2}} ds$$

$$= |x_0 - \bar{x}_0| + (a^2 + b^2)^{\frac{1}{2}} \int_0^t \|u(s) - v(s)\|\, ds$$

$$\leq |x_0 - \bar{x}_0| + (a^2 + b^2)^{\frac{1}{2}} \left(\int_0^t 1^2\, ds\right)^{\frac{1}{2}} \left(\int_0^t \|u(s) - v(s)\|^2\, ds\right)^{\frac{1}{2}}$$

$$= |x_0 - \bar{x}_0| + (a^2 + b^2)^{\frac{1}{2}} \sqrt{t} \left(\int_0^t \|u(s) - v(s)\|^2\, ds\right)^{\frac{1}{2}}.$$

Using of the inequality $(\alpha + \beta)^2 \leq 2(\alpha^2 + \beta^2)$ yields

$$|x(t) - \bar{x}(t)|^2 \leq 2\left(|x_0 - \bar{x}_0|^2 + (a^2 + b^2)T \int_0^t \|u(s) - v(s)\|^2\, ds\right).$$

Similarly,

$$|y(t) - \bar{y}(t)|^2 \leq 2\left(|y_0 - \bar{y}_0|^2 + (c^2 + d^2)T \int_0^t \|u(s) - v(s)\|^2\, ds\right).$$

Then

$$\|u(t) - v(t)\|^2 \leq 2\|u_0 - v_0\|^2 + 2(a^2 + b^2 + c^2 + d^2)T \int_0^t \|u(s) - v(s)\|^2\, ds.$$

Now Gronwall's inequality gives

$$\|u(t) - v(t)\|^2 \leq 2\|u_0 - v_0\|^2 e^{2T^2(a^2 + b^2 + c^2 + d^2)}.$$

Thus we have the estimation

$$\|u(t) - v(t)\| \leq \sqrt{2}\|u_0 - v_0\| e^{T^2(a^2 + b^2 + c^2 + d^2)}.$$

5.2 Proposed Exercises

Exercise 5.5. Find an upper bound for a function $f \in C[0, a]$ that satisfies on $[0, a]$ the inequality:
(a) $f(x) \leq 4 + 5 \int_0^x f(s)\,ds$;
(b) $f(x) \leq 4 + 5 \int_x^a f(s)\,ds$;
(c) $f(x) \leq -4 + 5 \int_0^x f(s)\,ds$.
Also give a direct proof in each of the three cases.

Exercise 5.6. Let $x, y \in C^1[0, T]$ be two solutions of the equation $x' = 2t^2 x + 3 \arctan x$. Estimate the difference $x(t) - y(t)$ in terms of the initial values $x(0) = x_0$ and $y(0) = y_0$.

Exercise 5.7. Prove that the solution $y \in C^1[0, h]$ of the Cauchy problem

$$\begin{cases} y' = t + \sin y \\ y(0) = 0 \end{cases}$$

differs from the solution of the linearized problem

$$\begin{cases} x' = t + x \\ x(0) = 0 \end{cases}$$

at most by $h(e^h - h - 1)^3 e^h / 6$. Give a generalization of this result.

Exercise 5.8. In the case of Exercise 5.4, give the estimate of $\|u(t) - v(t)\|$ in terms of another norm $\|\cdot\|$ in $\mathbb{R}^2$, for example: (a) $\|u\| = |x| + |y|$; (b) $\|u\| = \max\{|x|, |y|\}$.

5.3 Hints and Solutions

Exercise 5.5
(a) $f(x) \leq 4 \exp(5x)$;
(b) $f(x) \leq 4 \exp[5(a - x)]$;
(c) $f(x) \leq -4 \exp(5x)$.

Exercise 5.6
$|x(t) - y(t)| \leq |x_0 - y_0| \exp[(2T^2 + 3)T]$.

Exercise 5.7
Use the result from Exercise 5.3. Here,

$$f(t, x) = t + x, \quad g(t, x) = t + \sin x, \quad L = 1.$$

One has $x(t) = e^t - t - 1$, and from $\sin x \approx x$ with error $R \leq |x|^3/6$, we find

$$\eta = \max_{t \in [0,h]} |f(t, x(t)) - g(t, x(t))| = \max_{t \in [0,h]} |x(t) - \sin x(t)|$$

$$\leq \max_{t \in [0,h]} \frac{x(t)^3}{6} = \frac{(e^h - h - 1)^3}{6}.$$

5.4 Project: Integral and Differential Inequalities

Suggested structure. 1. Generalizations of Gronwall's inequality; 2. Inequalities of Harnack[1] type.

Basic bibliography. V. Barbu [2]; I. A. Rus [22, pp. 84–85]; A. Buică, *Gronwall-type nonlinear integral inequalities*, Mathematica 44 (2002), no. 1, 19–23; S. S. Dragomir, *Some Gronwall Type Inequalities and Applications*, Nova Science, 2003; S. Berhanu and A. Mohammed, *A Harnack inequality for ordinary differential equations*, Amer. Math. Montly, 112 (2005), no. 1, 32–41; D.-R. Herlea, *Positive solutions for second-order boundary-value problems with φ-Laplacian*, Electron. J. Differential Equations, 2016 (2016), no. 51, 1–8.

Hints. 1. (a) If φ is a continuous function on the interval $[t_0, t_1]$, a is a real number, b is a nonnegative continuous function on $[t_0, t_1]$, and

$$\varphi(t) \leq a + \int_{t_0}^{t} b(s)\varphi(s)\,ds, \quad \text{for all } t \in [t_0, t_1],$$

then

$$\varphi(t) \leq a \exp\left(\int_{t_0}^{t} b(s)\,ds\right), \quad \text{for all } t \in [t_0, t_1].$$

Example (*Boundedness of solutions*): Consider the differential system on semiline

$$x' = f(t, x), \quad t \in \mathbb{R}_+,$$

where $f: \mathbb{R}_+ \times \mathbb{R}^n \to \mathbb{R}^n$ is continuous. Assume that there are continuous functions $L, M: \mathbb{R}_+ \times \mathbb{R}_+ \to \mathbb{R}_+$ such that the following conditions are satisfied:
(i) $\|f(t, x)\| \leq L(t, \|x\|)$ for all $t \in \mathbb{R}_+$ and $x \in \mathbb{R}^n$;
(ii) $0 \leq L(t, u) - L(t, v) \leq (u - v)M(t, v)$ for all $t, u, v \in \mathbb{R}_+$ with $v \leq u$;
(iii) $\int_0^{+\infty} L(s, u)\,ds < \infty$ and $\int_0^{+\infty} M(s, u)\,ds < \infty$ for every $u \in \mathbb{R}_+$.

Then any solution x of the system that is defined on a semiline $[t_0, +\infty)$, where $t_0 \geq 0$, is bounded on $[t_0, +\infty)$.

---

[1] Carl Gustav Axel von Harnack (1851–1888)

Solution (sketch). One has

$$\|x(t)\| \leq \|x(t_0)\| + \int_{t_0}^{t} \|f(s, x(s))\| \, ds$$

$$\leq \|x(t_0)\| + \int_{t_0}^{t} L(s, \|x(s)\|) \, ds$$

$$\leq \|x(t_0)\| + \int_{t_0}^{t} (L(s, \|x(s)\| + \|x(t_0)\|) - L(s, \|x(t_0)\|)) \, ds$$

$$+ \int_{t_0}^{t} L(s, \|x(t_0)\|) \, ds \, .$$

Then

$$\|x(t)\| \leq \|x(t_0)\| + + \int_{t_0}^{+\infty} L(s, \|x(t_0)\|) \, ds + \int_{t_0}^{t} \|x(s)\| M(s, \|x(t_0)\|) \, ds \, .$$

The conclusion now follows from the generalized Gronwall's inequality from above, where

$$a = \|x(t_0)\| + + \int_{t_0}^{+\infty} L(s, \|x(t_0)\|) \, ds \quad \text{and} \quad b(t) = M(s, \|x(t_0)\|) \, .$$

(b) *Bihari[2]–LaSalle[3] inequality*: Let φ and b be nonnegative continuous functions on the interval $[t_0, t_1]$, $a \geq 0$, and $h: \mathbb{R}_+ \to \mathbb{R}_+$ a nondecreasing continuous function with $h(u) > 0$ for every $u > a$. Let

$$H(u) := \int_{a^+}^{u} \frac{ds}{h(s)} \quad \text{for } u > a \, ,$$

and assume either (i) $H(a) := \lim_{u \downarrow a} H(u) = 0$ and $H(+\infty) = \lim_{u \to +\infty} H(u) = +\infty$ (typical examples: $h(u) = u$ and $a > 0$; $h(u) = \sqrt{u}$), or (ii) $H(u) = +\infty$ for some (equivalently, for every) $u \geq a$ (typical example: $h(u) = u$ and $a = 0$).

If

$$\varphi(t) \leq a + \int_{t_0}^{t} b(s) h(\varphi(s)) \, ds, \quad \text{for all } t \in [t_0, t_1] \, ,$$

---

[2] Imre Bihari (1915–1998)
[3] Joseph P. LaSalle (1916–1983)

then, in case (i),

$$\varphi(t) \le H^{-1}\left(\int_{t_0}^{t} b(s)\,ds\right), \quad \text{for all } t \in [t_0, t_1], \tag{5.1}$$

while in case (ii),

$$\varphi(t) \le a, \quad \text{for all } t \in [t_0, t_1].$$

Proof: Let $\psi(t) = a + \int_{t_0}^{t} b(s)h(\varphi(s))\,ds$. For $t \in [t_0, t_1]$, one has

$$0 \le \varphi(t) \le \psi(t), \quad \psi'(t) = b(t)h(\varphi(t)).$$

Then

$$\psi'(t) \le b(t)h(\psi(t)).$$

Clearly, ψ is nondecreasing and $\psi(t) \ge a = \psi(t_0)$. If $\psi(t) = a$ for every $t \in [t_0, t_1]$, then the result is proved in both cases. Otherwise, there exists $t_2 \in [t_0, t_1)$ such that $\psi(t) = a$ on $[t_0, t_2]$ and $\psi(t) > a$ on $(t_2, t_1]$. Then on $(t_2, t_1]$, $h(\psi(t)) > 0$ and

$$\frac{\psi'(t)}{h(\psi(t))} \le b(t).$$

Integrating from t_2 to t yields

$$\int_{a_+}^{\psi(t)} \frac{ds}{h(s)} \le \int_{t_2}^{t} b(s)\,ds \le \int_{t_0}^{t} b(s)\,ds \quad (t_2 < t \le t_1),$$

that is

$$H(\psi(t)) \le \int_{t_0}^{t} b(s)\,ds.$$

Such an inequality cannot hold in the case (ii), while in the first case yields

$$\psi(t) \le H^{-1}\left(\int_{t_0}^{t} b(s)\,ds\right),$$

which together with $\varphi(t) \le \psi(t)$ gives (5.1) for $t \in [t_2, t_1]$. Finally, for $t \in [t_0, t_2]$, since $H^{-1}(\mathbb{R}_+) = [a, +\infty)$, one has

$$\varphi(t) \le \psi(t) = a \le H^{-1}\left(\int_{t_0}^{t} b(s)\,ds\right).$$

Example: If x solves the equation $x' = \sin\sqrt{tx}$ on $[0, T]$ and $x(0) = 0$, then

$$0 \le x(t) \le \frac{t^3}{9}, \quad t \in [0, T].$$

Indeed, one has

$$0 \le x(t) = \int_0^t \sin\sqrt{sx(s)}\,ds \le \int_0^t \sqrt{s}\sqrt{x(s)}\,ds.$$

Now use the Bihari–LaSalle inequality with $a = 0$, $b(t) = \sqrt{t}$ and $h(u) = \sqrt{u}$. Direct calculation gives $H^{-1}(\int_0^t b(s)\,ds) = t^3/9$.

2. In Section I.3.6 we considered the boundary value problem

$$\begin{cases} x'' + ax' + bx = h(t), & t \in [0, 1] \\ \alpha_1 x(0) + \alpha_2 x'(0) = 0 \\ \beta_1 x(1) + \beta_2 x'(1) = 0, \end{cases} \quad (5.2)$$

where $a, b \in \mathbb{R}$, $h \in C[0, 1]$ and $\alpha_1, \alpha_2, \beta_1, \beta_2$ are constants such that the zero function is the unique solution of the problem for $h = 0$. The solution of (5.2) can be expressed in the integral form

$$x(t) = -\int_0^1 G(t, s) h(s)\,ds,$$

where $G(t, s)$ is the Green's function

$$G(t, s) = \begin{cases} -\frac{\phi(t)\psi(s)}{W(s)}, & 0 \le t \le s \le 1 \\ -\frac{\phi(s)\psi(t)}{W(s)}, & 0 \le s < t \le 1. \end{cases}$$

Here ϕ and ψ are nonzero solutions of the equation $x'' + ax' + bx = 0$, which satisfy the first and the second boundary condition respectively.

Assume in addition that

$W(s)$ is negative on $[0, 1]$,

ϕ is positive nondecreasing on $(0, 1)$, and

ψ is positive nonincreasing on $(0, 1)$.

Under these conditions, one can easily check that for each subinterval $[\gamma, \delta] \subset (0, 1)$, Green's function has the properties:

$$0 \le G(t, s) \le G(s, s), \quad \text{for all } t, s \in [0, 1],$$
$$MG(s, s) \le G(t, s), \quad \text{for all } t \in [\gamma, \delta],\, s \in [0, 1],$$

where $M = \min\{\phi(\gamma)/\phi(1), \psi(\delta)/\psi(0)\}$.

Then, for any $x \in C^2[0, 1]$ satisfying the differential inequality

$$x'' + ax' + bx \le 0, \quad t \in [0, 1], \quad (5.3)$$

one has $x \geq 0$ on $[0, 1]$, and the following Harnack type inequality holds:
$$\min_{t \in [\gamma, \delta]} x(t) \geq M \|x\|_\infty, \tag{5.4}$$
where $\|x\|_\infty = \max_{t \in [0,1]} |x(t)|$. Indeed, if x solves (5.3) and we denote $g(t) := -(x''(t) + ax'(t) + bx(t))$, then $g \geq 0$ and
$$x(t) = \int_0^1 G(t, s) g(s) \, ds \ .$$
Clearly $x \geq 0$ on $[0, 1]$. Let $t^* \in [0, 1]$ be such that $x(t^*) = \|x\|_\infty$. Then, for any $t \in [\gamma, \delta]$, one has
$$x(t) = \int_0^1 G(t, s) g(s) \, ds \geq M \int_0^1 G(s, s) g(s) \, ds$$
$$\geq M \int_0^1 G(t^*, s) g(s) \, ds = M x(t^*) ,$$
which immediately gives (5.4).

Example (*Solution localization*): Consider the nonlinear boundary value problem
$$\begin{cases} x'' + ax' + bx + f(x) = 0, & t \in [0, 1] \\ \alpha_1 x(0) + \alpha_2 x'(0) = 0 \\ \beta_1 x(1) + \beta_2 x'(1) = 0 , \end{cases} \tag{5.5}$$
where $a, b, \alpha_1, \alpha_2, \beta_1, \beta_2$ are as above and $f : \mathbb{R}_+ \to \mathbb{R}_+$ is a nondecreasing continuous function. Let $[\gamma, \delta] \subset (0, 1)$ be a given interval, $M > 0$ be the corresponding constant from the Harnack type inequality, and $t_0 \in [0, 1]$ be a fixed number. If for two numbers $0 < r < R$, the inequality
$$\frac{f(Mr)}{R} > \frac{1}{\int_\gamma^\delta G(t_0, s) \, ds} \tag{5.6}$$
holds, then (5.5) has no nonnegative solutions with $r \leq \|x\|_\infty \leq R$.

To prove this result, assume the contrary: that there exists a nonnegative solution x with $r \leq \|x\|_\infty \leq R$. Then
$$R \geq \|x\|_\infty \geq x(t_0) = \int_0^1 G(t_0, s) f(x(s)) \, ds \geq \int_\gamma^\delta G(t_0, s) f(x(s)) \, ds$$
$$\geq \int_\gamma^\delta G(t_0, s) f(M \|x\|_\infty) \, ds \geq \int_\gamma^\delta G(t_0, s) f(Mr) \, ds ,$$
which contradicts (5.6).

Seminar 6 Method of Successive Approximations

6.1 Solved Exercises

Exercise 6.1. Write down the recurrence formula defining the sequence of successive approximations for the Cauchy problem

$$\begin{cases} x' = tx^2 + 2 \\ x(0) = 1, \end{cases}$$

and calculate the first three approximations of the solution.

Solution. The Cauchy problem is equivalent to the integral equation

$$x(t) = 1 + \int_0^t (sx(s)^2 + 2)\, ds,$$

or more explicitly

$$x(t) = 1 + 2t + \int_0^t sx(s)^2\, ds.$$

Then the sequence of successive approximations is defined as follows:

$$x_0(t) \equiv 1,$$

$$x_{k+1}(t) = 1 + 2t + \int_0^t sx_k(s)^2\, ds, \quad k = 0, 1, \dots.$$

The second approximation of the solution is $x_1(t) = 1 + 2t + t^2/2$, and the third one is

$$x_2(t) = 1 + 2t + \int_0^t sx_1(s)^2\, ds$$

$$= 1 + 2t + \int_0^t s\left(1 + 2s + \frac{s^2}{2}\right)^2 ds$$

$$= \frac{t^6}{24} + \frac{2t^5}{5} + \frac{5t^4}{4} + \frac{4t^3}{3} + \frac{t^2}{2} + 2t + 1.$$

Exercise 6.2. Define the sequence of successive approximations and find its limit for the Cauchy problem

$$\begin{cases} x' = 2x + 3, \quad t \in [0, 1] \\ x(0) = -\frac{1}{2}. \end{cases}$$

Solution. The equivalent integral equation is

$$x(t) = -\frac{1}{2} + \int_0^t (2x(s) + 3)\, ds.$$

Hence the sequence of successive approximations is

$$x_0(t) \equiv -\frac{1}{2},$$

$$x_{k+1}(t) = -\frac{1}{2} + 3t + 2\int_0^t x_k(s)\, ds, \quad k = 0, 1, \ldots.$$

Its limit is the solution of the Cauchy problem, which can be obtained by direct solving since the differential equation is linear. The solution is $x(t) = e^{2t} - 3/2$.

Exercise 6.3. Determine an interval $J_h = [-h, h]$ on which the sequence of successive approximations is well defined and uniformly convergent for the Cauchy problem

$$\begin{cases} x' = (4 - x^2)^{-\frac{1}{2}} + t, & t \in [-1, 1] \\ x(0) = 0. \end{cases}$$

Solution. The function $f(t, x) = (4 - x^2)^{-1/2} + t$ is well defined and continuous on $[-1, 1] \times (-2, 2)$. We choose any number $b < 2$, close to 2, and we consider $\Delta := [-1, 1] \times [-b, b]$. It is easy to see that on Δ, the function f is Lipschitz continuous in x. Let $h := \min\{1, b/M\}$. Here $M = \max_\Delta |f(t, x)| = (4 - b^2)^{-\frac{1}{2}} + 1$. Then $h = b\sqrt{4 - b^2}/(\sqrt{4 - b^2} + 1)$.

6.2 Proposed Exercises

Exercise 6.4. Define the sequence of successive approximations and find the first three approximations of the solution of the Cauchy problem

(a) $\begin{cases} x' = \sqrt{x} + 2t \quad (t \geq 0) \\ x(0) = 0; \end{cases}$

(b) $\begin{cases} x' = 2x^2 - t \\ x(1) = 0. \end{cases}$

Exercise 6.5. Define the sequence of successive approximations and find its limit for the Cauchy problem

(a) $\begin{cases} x' = 2x + 3, \quad t \in [0, 1] \\ x(0) = -\frac{1}{2}; \end{cases}$

(b) $\begin{cases} x' = \frac{1}{t}x + x^2, \quad t \in \left[1, \frac{15}{14}\right] \\ x(1) = 2. \end{cases}$

Exercise 6.6. Determine an interval $J_h = [-h, h]$ on which the sequence of successive approximations is well defined and uniformly convergent for the Cauchy problem

$$\begin{cases} x' = t + \arccos \frac{x}{\lambda}, & t \in [-1, 1] \\ x(0) = 0, \end{cases}$$

where $\lambda > 0$.

6.3 Hints and Solutions

Exercise 6.4
(a) $x_0(t) \equiv 0$, $x_{k+1}(t) = t^2 + \int_0^t \sqrt{x_k(s)}\, ds$, $k = 0, 1, \ldots$. The first three approximations: 0, t^2, $3t^2/2$.
(b) $x_0(t) \equiv 0$, $x_{k+1}(t) = -t^2/2 + 2\int_1^t x_k(s)^2\, ds$, $k = 0, 1, \ldots$. The first three approximations: 0, $-t^2/2$, $t^5/10 - t^2/2 - 1/10$.

Exercise 6.5
(a) $x_0(t) \equiv -1/2$, $x_{k+1}(t) = 3t + 2\int_0^t x_k(s)\, ds$, $k = 0, 1, \ldots$. The limit: $x = e^{2t} - 3/2$.
(b) $x_0(t) \equiv 2$, $x_{k+1}(t) = 2 + \int_1^t \left(\frac{1}{s}x_k(s) + x_k(s)^2\right) ds$, $k = 0, 1, \ldots$. The limit: $x = 2t/(2 - t^2)$.

Exercise 6.6
$a = 1$, $b = \lambda$, $M = 1 + \pi$, $h = \min\{1, \lambda/(1 + \pi)\}$.

6.4 Project: The Vectorial Method for the Treatment of Nonlinear Differential Systems

Suggested structure. 1. Vector version of Gronwall's inequality; 2. Componentwise localization of solutions for the Cauchy problem; 3. Componentwise localization of solutions for boundary value problems; 4. Matrices with spectral radius less than one.

Basic bibliography. R. Precup, *The role of matrices that are convergent to zero in the study of semilinear operator systems*, Math. Comp. Modelling **49** (2009), 703–708; R. Precup [20, Section 10.1].

Hints. In the common setting, as shown in Section I.4.5, the systems are seen as generalized equations where $\mathbb{R}^n$ replaces $\mathbb{R}$, and the Euclidian norm $\|\cdot\|$ of $\mathbb{R}^n$ (or any other equivalent norm) is used instead of the absolute value $|\cdot|$. A much more appropriate approach to systems is to consider properties that need not be uniform over the components. For instance, the localization of solutions may be realized componentwise using vector-valued norms and matrices instead of constants in Lipschitz and growth conditions. Under this approach, the systems appear as particular equations having the splitting property, and a larger diversity of results may be expected. Historically, it was Perov[1] who first explained the advantage of the vectorial approach to systems.

1. If $\Phi \in C(J, \mathbb{R}_+^n)$, $t_0 \in J$, $A \in \mathbb{R}_+^n$, $B \in \mathcal{M}_n(\mathbb{R}_+)$ and $\Phi(t) \leq A + B \int_{t_0}^{t} \Phi(s)\, ds$ for all $t \in [t_0, +\infty) \cap J$, then

$$\Phi(t) \leq e^{(t-t_0)B} A$$

for every $t \in [t_0, +\infty) \cap J$.

2. In what follows, the symbol $|\cdot|$ will be used to denote the vector-valued norm in $\mathbb{R}^n$ defined by

$$|x| = [|x_1|, |x_2|, \ldots, |x_n|]^T,$$

for $x = [x_1, x_2, \ldots, x_n]^T \in \mathbb{R}^n$.

Consider the Cauchy problem for an n-dimensional system

$$x' = f(t, x), \quad x(t_0) = x^0,$$

where $x = [x_1, x_2, \ldots, x_n]^T$, $x^0 = [x_1^0, x_2^0, \ldots, x_n^0]^T$ and $f = [f_1, f_2, \ldots, f_n]^T$.

We are interested in localizing a solution $x = [x_1, x_2, \ldots, x_n]^T$ componentwise, as follows:

$$\left|x_i(t) - x_i^0\right| \leq b_i, \quad i = 1, 2, \ldots, n.$$

If $b = [b_1, b_2, \ldots, b_n]^T$, then the above localization condition can be written in the vector form $|x(t) - x^0| \leq b$.

Assume that the functions f_i are defined, continuous, and Lipschitz continuous in x on the set

$$D = [t_0 - a, t_0 + a] \times \left[x_1^0 - b_1, x_1^0 + b_1\right] \times \ldots \times \left[x_n^0 - b_n, x_n^0 + b_n\right],$$

that is, there exist constants $l_{ij} \in \mathbb{R}_+$ for $i, j = 1, 2, \ldots, n$, such that

$$|f_i(t, x_1, x_2, \ldots, x_n) - f_i(t, y_1, y_2, \ldots, y_n)|$$
$$\leq l_{i1}|x_1 - y_1| + l_{i2}|x_2 - y_2| + \ldots + l_{in}|x_n - y_n|, \quad i = 1, 2, \ldots, n,$$

for every $(t, x_1, x_2, \ldots, x_n), (t, y_1, y_2, \ldots, y_n) \in D$.

It is easy to see that the sequence of successive approximations (x^k),

$$x^{k+1}(t) = x^0 + \int_{t_0}^{t} f(s, x^k(s))\, ds,$$

is well defined on the interval $J_h = [t_0 - h, t_0 + h]$, where

$$h = \min\left\{a, \frac{b_1}{M_1}, \ldots, \frac{bn}{M_n}\right\}$$

---

[1] Anatoliy Ivanovich Perov (1933–)

and $M_i = \max_D |f_i|$, and that the following vector estimate holds as in the case of a single equation:

$$|x^{k+1}(t) - x^k(t)| \le \frac{h^k}{k!} L^k N,$$

for $k = 0, 1, \ldots$, where L is the square matrix $[l_{ij}]_{i,j=1,2,\ldots,n}$, and $N = [N_1, N_2, \ldots, N_n]^T$ is such that $|x^1(t) - x^0| \le N$ componentwise for every $t \in J_h$. Furthermore, as in the scalar case (Section I.4.8.3), we may conclude about the convergence of the sequence of successive approximations to the unique solution of the Cauchy problem.

3. Consider the boundary value problem for a semilinear second-order differential system

$$\begin{cases} x'' + ax' + bx = f(t, x), \quad t \in [0, 1] \\ \alpha_1 x(0) + \alpha_2 x'(0) = 0 \\ \beta_1 x(1) + \beta_2 x'(1) = 0, \end{cases}$$

where $a, b \in \mathbb{R}$; $f \in C([0, 1] \times \mathbb{R}^n, \mathbb{R}^n)$; $\alpha_1, \alpha_2, \beta_1, \beta_2$ are real coefficients; and the assumption made in Section I.3.6 is fulfilled.

Assume that f is Lipschitz continuous in x, i.e.

$$|f(t, x) - f(t, y)| \le L|x - y|$$

for all $(t, x), (t, y) \in [0, 1] \times \mathbb{R}^n$, where L is a square matrix of order n of nonnegative numbers, and, as above, $|\cdot|$ has the meaning of a vector-valued norm on $\mathbb{R}^n$. If L^k tends to the zero matrix as $k \to \infty$, then the boundary value problem has a unique solution $x \in C^2([0, 1], \mathbb{R}^n)$.

The boundary value problem is equivalent to the integral system

$$x(t) = -\int_0^1 G(t, s) f(s, x(s)) \, ds, \quad t \in [0, 1],$$

where G is the corresponding Green's function as defined in Section I.3.6. We shall obtain a solution of the integral system as a limit of the sequence of successive approximations,

$$x^{k+1}(t) = -\int_0^1 G(t, s) f(s, x^k(s)) \, ds \quad (k = 0, 1, \ldots), \tag{6.1}$$

where the first approximation x^0 is an arbitrary function from $C([0, 1], \mathbb{R}^n)$.

If we use the symbol $\|\cdot\|$ to denote the vector-valued norm on $C([0, 1], \mathbb{R}^n)$, i.e.

$$\|x\| = [\|x_1\|_\infty, \|x_2\|_\infty, \ldots, \|x_n\|_\infty]^T,$$

for $x = [x_1, x_2, \ldots, x_n]^T \in C([0,1], \mathbb{R}^n)$, where $\|x_i\|_\infty = \max_{t \in [0,1]} |x_i(t)|$, then we have the estimate

$$|x^{k+1}(t) - x^k(t)| \leq L \int_0^1 |G(t,s)| \, |x^k(s) - x^{k-1}(s)| \, ds$$

$$\leq ML\|x^k - x^{k-1}\|,$$

with $M = \max_{t \in [0,1]} \int_0^1 |G(t,s)| \, ds$. Taking the maximum with respect to t yields

$$\|x^{k+1} - x^k\| \leq ML\|x^k - x^{k-1}\|.$$

As a consequence,

$$\|x^{k+1} - x^k\| \leq ML^k\|x^1 - x^0\|.$$

This gives

$$\|x^{k+p} - x^k\| \leq M\left(L^k + L^{k+1} + \ldots + L^{k+p-1}\right)\|x^1 - x^0\|.$$

Since $L^k \to 0$ as $k \to \infty$, the matrix series $I + L + L^2 + \ldots$ is convergent. Consequently, the sequence (x^k) is Cauchy, so convergent to some function $x \in C([0,1], \mathbb{R}^n)$. Passing to limit in (6.1) yields the conclusion that x solves the integral system.

4. For a square matrix $L \in \mathcal{M}_n(\mathbb{R}_+)$, the following statements are equivalent:
(i) $L^k \to 0$ as $k \to \infty$.
(ii) $I - L$ is nonsingular and $(I - L)^{-1} = I + L + L^2 + \ldots$.
(iii) $\rho(L) < 1$, where $\rho(L)$ is the *spectral radius* of the matrix L, i.e. the maximum of absolute values of its eigenvalues.
(iv) $I - L$ is nonsingular and $(I - L)^{-1}$ has nonnegative entries.

To show that (i) implies (ii), we first prove that $I - L$ is nonsingular. For this it suffices to show that the unique solution of the homogeneous system of linear equations $(I - L)x = 0$ is zero. Indeed, if x is a solution, then $x = Lx$, so that we successively find $x = Lx = L^2x$, $x = L^3x$, and in general $x = L^k x$ for every k. Letting $k \to +\infty$ yields $x = 0$, as desired. Next, from

$$I - (I - L)\left(I + L + \ldots + L^k\right) = L^{k+1},$$

we have

$$I + L + \ldots + L^k = (I - L)^{-1}(I - L^{k+1}) \qquad (6.2)$$

and taking the limit gives the conclusion. The implication (ii) to (i) is immediate since the convergence of the series implies that its general term tends to zero.

The equivalence of (i) and (iii) can be established using the Jordan canonical representation $L = P^{-1}JP$. It is easy to see that $L^k = P^{-1}J^kP$ for every k, and that J^k is a diagonal matrix for $k \geq n$, whose diagonal entries are the k-th powers of the eigenvalues of L. Hence $L^k \to 0$ is equivalent to $J^k \to 0$, that is, to $r^k \to 0$ for all eigenvalues r of L.

6.4 Project: The Vectorial Method for the Treatment of Nonlinear Differential Systems

Clearly (ii) implies (iv). Finally to prove that (iv) implies (ii), use again the identity (6.2), which can be written as

$$I + L + \ldots + L^k = (I - L)^{-1} - (I - L)^{-1} L^{k+1}. \tag{6.3}$$

Since both matrices $(I-L)^{-1}$ and L^{k+1} have nonnegative entries, the right-hand side is dominated by $(I-L)^{-1}$. Hence the partial sums of the series $I + L + L^2 + \ldots$ are bounded. Consequently the series is convergent and letting $k \to +\infty$ in (6.3) finishes the proof.

Seminar 7 Stability of Solutions

7.1 Solved Exercises

Exercise 7.1. Discuss the stability of the system

$$\begin{cases} x' = ax - y \\ y' = 3x + y \end{cases}$$

in terms of real parameter a.

Solution. The system is asymptotically stable if and only if tr $A < 0$ and det $A > 0$. One has tr $A = a + 1$ and det $A = a + 3$. Hence the system is asymptotically stable for $a \in (-3, -1)$. For $a = -1$ and $a = -3$, the system is stable, and for $a \in (-\infty, -3) \cup (-1, +\infty)$ it is unstable.

Exercise 7.2. Show that the nonlinear system

$$\begin{cases} x' = -x - 5y^2 \\ y' = xy - 2y^2 \end{cases}$$

has an asymptotically stable stationary solution.

Solution. To find the stationary solutions we have to solve the algebraic system $-x - 5y^2 = 0$, $xy - 2y^2 = 0$. To this aim, multiply by -2 the first equation, by 5 the second one, and then add them. One obtains $5xy + 2x = 0$, where $x = 0$, or $y = -2/5$. For $x = 0$, we have $y = 0$, while $y = -2/5$ yields $x = -4/5$. Hence the stationary solutions are $(0, 0)$ and $(-4/5, -2/5)$. The Jacobian matrix of the system is

$$\begin{bmatrix} -1 & -10y \\ y & x - 4y \end{bmatrix}.$$

It is Hurwitzian only for the second stationary solution. Therefore $(-4/5, -2/5)$ is an asymptotically stable stationary solution.

Exercise 7.3. Consider the following two nonlinear systems:

(a) $\begin{cases} x' = -x(x^2 + y^2) - y \\ y' = -y(x^2 + y^2) + x; \end{cases}$

(b) $\begin{cases} x' = x(x^2 + y^2) - y \\ y' = y(x^2 + y^2) + x. \end{cases}$

Prove that the null solution is asymptotically stable for system (a), and unstable for system (b).

Solution. In both cases, the Jacobian matrix for the null solution is not Hurwitzian since its eigenvalues are $\pm i$. However, in the first case, from

$$xx' + yy' = -(x^2 + y^2)^2$$

we deduce that $x^2 + y^2 = 1/(2t + C)$, where $1/C = x(0)^2 + y(0)^2$. This shows that the saturated solutions are defined on the whole positive semiline and $x^2 + y^2 \to 0$ as $t \to +\infty$. Hence the null solution is asymptotically stable. For the second system, one has $xx' + yy' = (x^2 + y^2)^2$, where $x^2 + y^2 = 1/(C - 2t)$. Hence the saturated solutions are defined only on the finite interval $[0, C/2)$ and blow up in finite time. Therefore the null equilibrium is unstable.

This example shows that the simple stability of the linearized system says nothing about the stability of the equilibrium of the nonlinear system.

Exercise 7.4. Use Lyapunov's direct method to prove the asymptotic stability of the null solution of the system

$$\begin{cases} x' = x^3 - 3x + y \\ y' = -x - y. \end{cases}$$

Solution. Observe that $x'x + y'y = x^4 - 3x^2 - y^2 < 0$ for $-\sqrt{3} < x < \sqrt{3}$ and $(x, y) \neq (0, 0)$. Hence $V(x, y) = x^2 + y^2$ is a Lyapunov function of the system on the set $G = \{(x, y): -\sqrt{3} < x < \sqrt{3}\}$, which satisfies the strict inequality (I.5.18). Therefore the null solution is asymptotically stable.

Exercise 7.5. Consider the *Lorenz*[1] system

$$\begin{cases} x' = a(y - x) \\ y' = bx - y - xz \\ z' = xy - cz, \end{cases}$$

where $a, b, c > 0$. Use Lyapunov's direct method to prove that for $b < 1$ the null solution is asymptotically stable. Moreover, prove the system is globally asymptotically stable.

Solution. Observe that

$$x'x + ay'y + az'z = -a\left[x^2 + y^2 - (1 + b)xy + cz^2\right].$$

If $b < 1$, then $x^2 + y^2 - (1 + b)xy > 0$ for $x^2 + y^2 \neq 0$, hence $-a[x^2 + y^2 - (1 + b)xy + cz^2] < 0$ for every $(x, y, z) \neq (0, 0, 0)$. Therefore $V(x, y, z) = x^2 + ay^2 + az^2$ is a Lyapunov function that satisfies the condition (I.5.18). Hence the null solution is asymptotically stable. Moreover, the assumptions of Theorem I.5.28 are fulfilled, hence the system is globally asymptotically stable.

---

[1] Edward Norton Lorenz (1917–2008)

Notice that for $b > 1$, the null solution is unstable and the behavior of the orbits is chaotic (e.g. [12, Section 14.5], [13, p. 249] and [26, Section 5.7]).

7.2 Proposed Exercises

Exercise 7.6. Check for stability the following linear systems:

(a) $\begin{cases} x' = x - 2y \\ y' = 3x + y; \end{cases}$

(b) $\begin{cases} x' = -5x + ay \\ y' = x + y; \end{cases}$

(c) $\begin{cases} x' = y + z \\ y' = z + x \\ z' = x + y; \end{cases}$

(d) $\begin{cases} x' = z - y \\ y' = x - z \\ z' = y - x. \end{cases}$

Exercise 7.7. Show that each of the following nonlinear systems:

(a) $\begin{cases} x' = -x + y^2 - 1 \\ y' = x; \end{cases}$

(b) $\begin{cases} x' = 2x(y^2 - 1) - y \\ y' = x, \end{cases}$

has an asymptotically stable stationary solution.

Exercise 7.8. Check for stability the null solution of the following nonlinear systems:

(a) $\begin{cases} x' = xy - x \\ y' = -x^2 - 2y; \end{cases}$

(b) $\begin{cases} x' = -x + 2\sin y \\ y' = -3\tan x - 5y. \end{cases}$

Exercise 7.9. Use Lyapunov's direct method to prove the asymptotic stability of the null solution of the system

$$\begin{cases} x' = y - xf(x, y) \\ y' = -x - yg(x, y), \end{cases}$$

under the assumption that $f(x, y) > 0$ and $g(x, y) > 0$ for $(x, y) \neq (0, 0)$ in some neighborhood of the origin.

Exercise 7.10. Use Lyapunov's direct method to prove the asymptotic stability of the null solution of the system

$$\begin{cases} x' = -2y + yz - x^3 \\ y' = x - xz - y^3 \\ z' = xy - z^3 \, . \end{cases}$$

7.3 Hints and Solutions

7.6

(a) Unstable; (b) asymptotically stable if $a < -5$, stable for $a = -5$, unstable if $a > -5$; (c) one root of the characteristic equation is bigger than 1, so the system is unstable; (d) the roots of the characteristic equation are 0 and $\pm i\sqrt{3}$. Since all these roots have the real part 0 and are simple, the system is stable.

7.7

(a) $(0, -1)$; (b) $(0, 0)$.

7.8

(a), (b) For $x = y = 0$, the Jacobian matrix is Hurwitzian, so the null solution is asymptotically stable.

7.9

The function $V(x, y) = x^2 + y^2$ is a Lyapunov function on that neighborhood of the origin, and the condition (I.5.18) is satisfied.

7.10

A Lyapunov function is $V(x, y, z) = x^2 + 2y^2 + z^2$.

7.4 Project: Stable and Unstable Invariant Manifolds

Suggested structure. 1. Invariant subspaces associated with linear differential systems; 2. Invariant manifolds associated with equilibria of nonlinear systems; 3. The stable manifold theorem.

Basic bibliography. L. Perko [18, pp. 104–118]

Hints. Consider the differential system $x' = f(x)$, where $f \in C^1(D, \mathbb{R}^n)$ and $D \subset \mathbb{R}^n$ is an open set. Assume that for each $a \in D$, the Cauchy problem with the initial condition $x(0) = a$ has a unique saturated solution $x(., a)$ defined on the whole real axis. The mapping $\phi \colon \mathbb{R} \times D \to D$ given by $\phi(t, a) = x(t, a)$ is called the *flow* of the differential system. For each $t \in \mathbb{R}$, denote by ϕ_t the mapping from D to D, defined by $\phi_t(a) = \phi(t, a)$. One has

$$\phi_{t+s} = \phi_t \phi_s, \quad \text{for all } t, s \in \mathbb{R}\,.$$

A set $M \subset D$ is said to be *invariant* with respect to the flow if $\phi_t(M) \subset M$ for every $t \in \mathbb{R}$, or equivalently if $\phi_t(M) = M$ for every $t \in \mathbb{R}$. Obviously, any orbit of the system is an invariant set. Additionally, D is an invariant set. We say that M is a *positively* (or *negatively*) *invariant set* if $\phi_t(M) \subset M$ for every $t \geq 0$ (or $t \leq 0$).

1. Consider the constant coefficient linear differential system $x' = Ax$, where A is a square matrix of order n. In this case $\phi_t(a) = e^{tA}a$. Denote

$$E^s = \{a \in \mathbb{R}^n : e^{tA}a \to 0 \text{ as } t \to +\infty\},$$
$$E^u = \{a \in \mathbb{R}^n : e^{tA}a \to 0 \text{ as } t \to -\infty\}.$$

It is easy to see that E^s and E^u are linear subspaces of $\mathbb{R}^n$ and that $E^s \cap E^u = \{0\}$. Then there is a subspace E^c of $\mathbb{R}^n$ such that

$$\mathbb{R}^n = E^s \oplus E^c \oplus E^u.$$

The spaces E^s, E^u and E^c are *invariant* with respect to the flow and are respectively called the *stable, unstable* and *center subspace* (manifold) of the linear differential system. Characterizations of these subspaces in terms of eigenvalues of matrix A are known ([18]). For example, if all eigenvalues of A are real and simple, then E^s and E^u are the linear subspaces of $\mathbb{R}^n$ generated by the eigenvectors corresponding to the negative and positive eigenvalues respectively.

Example: Find the stable and unstable subspaces (manifolds) of the linear system $x' = Ax$, in each of the following cases:

(a) $A = \begin{bmatrix} -1 & 1 \\ 3 & 1 \end{bmatrix}$;

(b) $A = \begin{bmatrix} -1 & 0 & 0 \\ 0 & -2 & 0 \\ 0 & 0 & 1 \end{bmatrix}$.

Solution. (a) If $a = [a_1, a_2]^T$, then

$$\phi_t(a) = \left[\frac{3a_1 - a_2}{4} e^{-2t} + \frac{a_1 + a_2}{4} e^{2t}, \ -\frac{3a_1 - a_2}{4} e^{-2t} + 3\frac{a_1 + a_2}{4} e^{2t} \right]^T.$$

Then $\phi_t(a) \to 0$ as $t \to +\infty$ only if $a_1 + a_2 = 0$. Hence

$$E^s = \{a \in \mathbb{R}^2 : a_1 + a_2 = 0\},$$

i.e. the line of equation $x_1 + x_2 = 0$ is the stable manifold of the system. Additionally, $\phi_t(a) \to 0$ as $t \to -\infty$, only if $3a_1 - a_2 = 0$. Hence

$$E^u = \{a \in \mathbb{R}^2 : 3a_1 - a_2 = 0\},$$

i.e. the unstable manifold of the system is the the line of equation $3x_1 - x_2 = 0$.

7.4 Project: Stable and Unstable Invariant Manifolds — 175

(b) If $a = [a_1, a_2, a_3]^T$, then
$$\phi_t(a) = \begin{bmatrix} a_1 e^{-t}, & a_2 e^{-2t}, & a_3 e^t \end{bmatrix}^T.$$

Hence $\phi_t(a) \to 0$ as $t \to +\infty$ only if $a_3 = 0$. Therefore, $E^s = \{a \in \mathbb{R}^3 : a_3 = 0\}$. Additionally, $\phi_t(a) \to 0$ as $t \to -\infty$ only if $a_1 = a_2 = 0$. So, $E^u = \{a \in \mathbb{R}^3 : a_1 = a_2 = 0\}$. Hence, the stable manifold of the system is the plane $x_3 = 0$, and the unstable manifold is the line $x_1 = x_2 = 0$.

2. Consider the (nonlinear) differential system $x' = f(x)$, where $f \in C^1(D, \mathbb{R}^n)$ and $D \subset \mathbb{R}^n$ is an open set. Assume that $f(x_0) = 0$, i.e. $x = x_0$ is an equilibrium solution of the system. If x_0 is a stable node, then $\phi_t(a) \to x_0$ as $t \to +\infty$ for every a in some neighborhood V of x_0. If x_0 is an unstable node, then $\phi_t(a) \to x_0$ as $t \to -\infty$ for every a in some neighborhood V of x_0. If none of these two situations holds, then locally for a given neighborhood V of x_0, we may consider the sets

$$M_V^s := \{a \in V : \phi_t(a) \in V \text{ for all } t \geq 0 \text{ and } \phi_t(a) \to x_0 \text{ as } t \to +\infty\},$$

$$M_V^u := \{a \in V : \phi_t(a) \in V \text{ for all } t \leq 0 \text{ and } \phi_t(a) \to x_0 \text{ as } t \to -\infty\}.$$

It is easy to see that M_V^s is a positively invariant set, while M_V^u is a negatively invariant set. In the case where M_V^s does not reduce to the singleton $\{x_0\}$, we call it the (local) *stable manifold* of x_0. Similarly, if $M_V^u \neq \{x_0\}$, M_V^u is said to be the (local) *unstable manifold* of x_0.

Let
$$M^s := \cup_{t \leq 0} \phi_t(M_V^s), \quad M^u := \cup_{t \geq 0} \phi_t(M_V^u).$$

We call M^s the *global stable manifold* of x_0, and M^u the *global unstable manifold* of x_0.
(a) The definition of M^s and M^u is independent on the neighborhood V of x_0.
(b) M^s and M^u are invariant sets with respect to the flow.
(c) For a linear system with constant coefficients $x' = Ax$ and $x_0 = 0$, one has

$$M^s = \text{span } M_V^s = E^s, \quad M^u = \text{span } M_V^u = E^u.$$

Example: Find the stable and unstable manifolds of the zero equilibrium of the nonlinear systems

(a) $\begin{cases} x_1' = -x_1 + x_2^2 \\ x_2' = 2x_2; \end{cases}$

(b) $\begin{cases} x_1' = -x_1 \\ x_2' = -2x_2 + x_1^2 \\ x_3' = x_3 - x_1^2. \end{cases}$

Solution. (a) Solving the system yields
$$\phi_t(a) = \left[\left(a_1 - \frac{a_2^2}{5} \right) e^{-t} + \frac{a_2^2}{5} e^{4t}, \; a_2 e^{2t} \right]^T.$$

We have $\phi_t(a) \to 0$ as $t \to +\infty$ if $a_2 = 0$, and $\phi_t(a) \to 0$ as $t \to -\infty$, for $a_1 - a_2^2/5 = 0$. Hence the stable manifold of the zero equilibrium is the axis $x_2 = 0$, and the unstable manifold is the parabola of equation $x_1 - x_2^2/5 = 0$.

(b) Solving the system gives

$$\phi_t(a) = \left[a_1 e^{-t},\ a_1^2 t e^{-t} + a_2 e^{-2t},\ \frac{a_1^2}{3} e^{-2t} + a_3 e^t \right]^T.$$

The stable manifold is the plane $x_3 = 0$, while the unstable manifold is the line $x_1 = x_2 = 0$.

3. The connection between the stable and unstable manifolds of a nonlinear system $x' = f(x)$ and the stable and unstable subspaces of its linearized system $x' = f_x(0)x$, where $f_x(0)$ is the Jacobian matrix of f at zero, is given by the following theorem:

Theorem (The stable manifold theorem). *Let $f \in C^p(D, \mathbb{R}^n)$ where D is an open subset of $\mathbb{R}^n$ containing the origin and $p \geq 1$. Assume that $f(0) = 0$ and that $f_x(0)$ has k eigenvalues with negative real part, j eigenvalues with positive real part, and $m = n - k - j$ eigenvalues with zero real part. Then the global stable manifold M^s of the origin is a k-dimensional manifold of class C^p, tangent at 0 to the stable subspace E^s of the linearized system, and the global unstable manifold M^u of the origin is a j-dimensional manifold of class C^p, tangent at 0 to the unstable subspace E^u of the linearized system. Additionally, there is an m-dimensional manifold M^c of class C^p tangent at 0 to the subspace E^c of the linearized system, called the center manifold of the origin.*

Part III: **Maple Code**

Lab 1 Introduction to Maple

Computers are today an essential, almost unavoidable tool for solving and analyzing different types of mathematical problems, particularly differential equations. Not long ago, solving problem described by an algorithm with a computer required the elaboration of a computer program. Now complex computation systems are available, such as MATLAB, Maple and Mathematica, which save us from individual programming of each problem or class of problems.

Maple is a computer algebra system of *symbolic computation*, which is able to give exact solutions in analytical form to numerous problems including limits of functions, derivatives, integrals, equations and systems of equations, linear algebra, differential equations, optimization problems, etc. In addition to *symbolic computation packages*, Maple incorporates many *numerical algorithms* for approximating solutions to problems that do not admit exact solutions. It also contains *graphic programs* allowing visualization of information and results.

Maple is structured on two levels. The basic level, the 'kernel', of about 10%, contains primitive and fundamental commands, and algorithms for arithmetic and polynomial calculus. The higher level, 'Maple's library', of about 90%, contains Maple code that can be called on command, and numerous specialized packages, such as LinearAlgebra for calculus with matrices, and DEtools for solving differential equations and visualizing their solutions.

The aim in this part of the book is to offer a short list of Maple commands that are useful for ordinary differential equations. Much more information and examples can be found in Maple tutorials and manuals[1] and more specialized textbooks such as [3, 7] and [17].

1.1 Numerical Calculus

- *Basic operations*: +, -, * (multiplication), / (division), ** or ^ (exponentiation), ! (factorial operation).

- *Functions*: sqrt (square root), exp (natural exponential function), ln (natural logarithm), log[10] (common logarithm), sin, cos, tan, cot, arcsin, arccos, arctan, sinh, cosh, tanh, trunc (integer part), frac (fractional part), abs (absolute value), max, min.

---

1   1. P. Adams, K. Smith and R. Výborný, *Introduction to Mathematics with Maple*, World Scientific, 2004;
    2. J.-M. Cornil and P. Testud, *An Introduction to Maple V*, Springer, 2001.
    3. http://www.maplesoft.com/support/help/Maple/view.aspx?path=HowDoI/ SolveAnOrdinaryDifferentialEquation

- *Standard constants:* Pi, e, I (imaginary unit $\sqrt{-1}$).
- *Brackets:* (,) – for order of operations; [,] – for a list; {,} – for a set.
- *Output display:* ; (semicolon). If instead of ';' one uses ':' (colon), then the command is taken into consideration but is not displayed, while for a calculation, the result is not displayed.
- *Display of results* is centered after pushing Enter.
- *Evaluation* of numerical expressions such as $\sqrt{2}$, π, or e^2, is done using the command evalf.

Here are some examples:
```
> 2 + 5 - 4;
```
$$3$$

```
> (3 + 7)*6 - 20;
```
$$40$$

```
> 2**4 + 3**2;
```
$$25$$

```
> 3/2 + 7/4;
```
$$\frac{13}{4}$$

```
> 1.5 + 7/4;
```
$$3.250000000$$

```
> sqrt(5);
```
$$\sqrt{5}$$

```
> evalf(sqrt(5));
```
$$2.236067977$$

```
> sqrt(5.0);
```
$$2.236067977$$

```
> exp(3);
```

$$e^3$$

```
> ln(exp(7));
```

$$7$$

```
> sin(Pi/4);
```

$$\tfrac{1}{2}\sqrt{2}$$

```
> (2 + 3*I) + (- 1 + 5*I);
```

$$1 + 8I$$

```
> abs(3-4*I);
```

$$5$$

1.2 Symbolic Calculus

Expressions:

All the functions from above admit variable arguments yielding to expressions. Here are some examples:
```
> ln(x**2 + y);
```

$$\ln(x^2 + y)$$

To compute the value of the expression for given values of the variables one makes the assignment, for example:
```
> x:=3; y:=4;
```

$$\ln(13)$$

If the numerical value is desired we use the command
```
> evalf(ln(x**2 + y));
```

$$2.564949357$$

To remove the assigned values use
```
> restart:
```

We give names to expressions, equations or systems of equations in the following way:

$$\text{name} := \text{expression};$$

Example 1: The commands

```
> E:=ln(x**2 + y);
> x:=3; y:=4;
```

have the output

$$E := \ln(13)$$

and the evaluation is obtained with the command

```
> evalf(E);
```

Example 2: To give the name Eq1 to the equation $x - 4y = 3$ we use
```
> Eq1:=x-4*y = 3;
```

$$Eq1 := x - 4y = 3$$

Functions – Plane curves

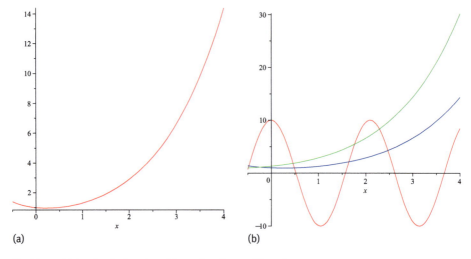

(a)   (b)

Fig. III.1.1: (a) Graph of a function; (b) graphs of several functions.

Define a function. Values of a function. Graph of a function:
```
> f:=x → 2**x-ln(x + 1);
```

$$f := x \to 2^x - \ln(x + 1)$$

> f(1), f(2);

$$2 - \ln(2), 4 - \ln(3)$$

To plot the graph of the function we must load the package {plots} using the command with(plots) and then the command plot (Figure III.1.1 (a)).

> with(plots):
> plot(f(x), x = -0.5..4);

Plot several graphs on the same figure (Figure III.1.1 (b)):

> plot({f(x), f(x+1), 10*cos(3*x)}, x = -0.5..4,
    color=[red,blue,green]);

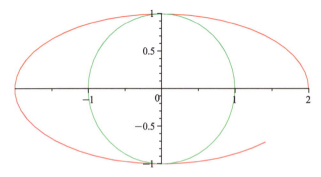

Fig. III.1.2: Representation of two parametrized curves.

For the graph representation of discontinuous functions, if we wish not to see the vertical asymptotes, use the option discont=true:

> plot(tan(x), x = -5*Pi/2..5*Pi/2, y = -4..4, discont=true);

Plot parametrized curves (to have the same scale on the two axes, use the option scaling=constrained):

> plot([2*cos(t),sin(t),t = 0..7*Pi/4], scaling=constrained);

Plot two parametrized curves (Figure III.1.2):

> plot([[2*cos(t), sin(t), t=0..7*Pi/4], [cos(t), sin(t),
    t=0..2*Pi]], scaling=constrained);

Plot a curve given by Cartesian equation using the command implicitplot:

> implicitplot(x**2+y**2=1, x=-1..1, y=-1..1);

To visualize the dependence of a curve on a parameter use `animate`, right-click on the image, select Animation and then Play:

```
> animate(sin(x*t), x=-Pi..Pi, t=1..3);
```

Functions of two variables

Definition:

```
> g:=(x,y) → sin(x*y);
```

Three-dimensional graphs using the command `plot3d`:

```
> plot3d(g(x,y), x=-2..2, y=-3..3);
```

Animation of three-dimensional graphs: use `animate3d`, right-click on the image, select Animation and then Play:

```
> animate3d(g(t*x,y), x=-2..2, y=-3..3, t=1..3);
```

Lab 2 Differential Equations with Maple

2.1 The DEtools Package

For solving and plotting differential equations the package DEtools must first be loaded:

```
> with(DEtools):
> eq:=diff(x(t),t)=1/2*x(t)-t;
```

$$eq := \frac{d}{dt}x(t) = \frac{1}{2}x(t) - t$$

Find the *general solution* of an ordinary differential equation. The output gives the solution in terms of arbitrary constants _C1, _C2, . . .:

```
> dsolve(eq,x(t));
```

$$x(t) = 4 + 2t + e^{\frac{1}{2}t}\_C1$$

Solve a Cauchy problem:

```
> ic:=x(0)=2;
```

$$ic := x(0) = 2$$

```
> dsolve({eq,ic},x(t));
```

$$x(t) = 4 + 2t - 2e^{\frac{1}{2}t}$$

Plot the solution of a Cauchy problem (Figure III.2.1 (a)):

```
> DEplot(eq,x(t),t=0..4,[[ic]],linecolor=red);
```

Plot the solution of a Cauchy problem *without representation of the vector (or direction) field* (Figure III.2.1 (b)):

```
> DEplot(eq,x(t),t=0..4,[[ic]],linecolor=red, arrows=none);
```

Plot *multiple solutions* corresponding to a list of initial conditions (Figure III.2.2 (a)):

```
> DEplot(eq,x(t),t=0..4,[x(0)=2,x(0)=2.5,x(0)=3],
    linecolor=[red, blue, green], arrows=none);
```

Draw the *direction field* of a differential equation (Figure III.2.2 (b)):

```
> dfieldplot(eq,x(t),t=0..4,x=-4..8);
```

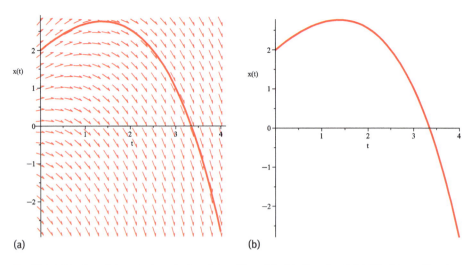

Fig. III.2.1: Plotting of the solution of a Cauchy problem, (a) with direction field; (b) without direction field.

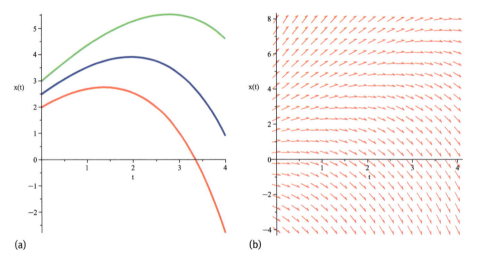

Fig. III.2.2: (a) Plotting of several solutions; (b) the direction field.

2.2 Working Themes

Theme 2.1. Solve with Maple the first-order differential equations from the Examples in Section I.1.2:
(a) Find the general solution;
(b) Find the solution of a Cauchy problem;
(c) Plot the solution of a Cauchy problem with and without representation of the vector field;

(d) Plot multiple solutions of an equation for a list of initial conditions;
(e) Draw the direction field of an equation.

Theme 2.2. Investigate with Maple the mathematical models from Section I.1.3 and give interpretations in terms of the modeled processes. Consider qualitative properties of the solutions such as: monotonicity, extrema, inflection points, asymptotic behavior, stability, and solution dependence on parameters.

Lab 3 Linear Differential Systems

3.1 The linalg Package

This package is used for definition of a matrix, eigenvalues, eigenvectors, inverse of a matrix, and solving the linear algebraic system $Ax = b$:

```
> with(linalg):
> A:=matrix(2,2,[[-2,3],[-3,-2]]);
```

$$A := \begin{bmatrix} -2 & 3 \\ -3 & -2 \end{bmatrix}$$

```
> eigenvalues(A);
```

$$-2 + 3I, \ -2 - 3I$$

```
> eigenvectors(A);
```

$$[-2 + 3I, \ 1, \ \{[-I \ 1]\}], \ [-2 - 3I, \ 1, \ \{[I \ 1]\}]$$

```
> inverse(A);
```

$$\begin{bmatrix} -\frac{2}{13} & -\frac{3}{13} \\ \frac{3}{13} & -\frac{2}{13} \end{bmatrix}$$

```
> b:=matrix(2,1,[4,3]);
```

$$b := \begin{bmatrix} 4 \\ 3 \end{bmatrix}$$

```
> x:=linsolve(A,b);
```

$$x := \begin{bmatrix} -\frac{17}{13} \\ \frac{6}{13} \end{bmatrix}$$

3.2 Linear Differential Systems

Solving linear differential systems:

```
> with(DEtools):
> with(plots):
> eq1:=diff(x(t),t)=-2*x(t)+3*y(t);
```

$$eq1 := \frac{d}{dt}x(t) = -2x(t) + 3y(t)$$

```
> eq2:=diff(y(t),t)=-3*x(t)-2*y(t);
```

$$eq2 := \frac{d}{dt}y(t) = -3x(t) - 2y(t)$$

3.2 Linear Differential Systems

```
> ds:=eq1,eq2;
```

$$ds := \frac{d}{dt}x(t) = -2x(t) + 3y(t), \frac{d}{dt}y(t) = -3x(t) - 2y(t)$$

```
> ic:=x(0)=-0.2,y(0)=0.5;
```

$$ic := x(0) = -0.2, y(0) = 0.5$$

```
> dsolve({ds},{x(t),y(t)});
```

$$\{x(t) = -e^{-2t}(\_C1\cos(3t) - \_C2\sin(3t)),$$
$$y(t) = e^{-2t}(\_C1\sin(3t) + \_C2\cos(3t))\}$$

```
> dsolve({ds,ic},{x(t),y(t)});
```

$$\left\{x(t) = -e^{-2t}\left(\frac{1}{5}\cos(3t) - \frac{1}{2}\sin(3t)\right),\right.$$
$$\left.y(t) = e^{-2t}\left(\frac{1}{5}\sin(3t) + \frac{1}{2}\cos(3t)\right)\right\}$$

The fundamental matrix $\exp(tA)$ of the linear system $u' = Au$:

```
> with(linalg):
> U:=exponential(A,t);
```

$$U := \begin{bmatrix} e^{-2t}\cos(3t) & e^{-2t}\sin(3t) \\ -e^{-2t}\sin(3t) & e^{-2t}\cos(3t) \end{bmatrix}$$

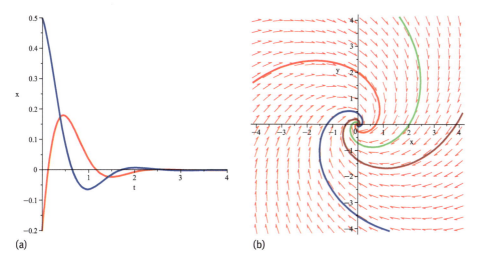

Fig. III.3.1: (a) Time series representation; (b) phasic portrait.

Plot a *time series* of the solution of a Cauchy problem (Figure III.3.1 (a)):

```
> X:=DEplot([ds],[x(t),y(t)],t=0..4,[[ic]],stepsize=0.01,
    linecolor=red,scene=[t,x]):
> Y:=DEplot([ds],[x(t),y(t)],t=0..4,[[ic]],stepsize=0.01,
    linecolor=blue,scene=[t,y]):
> display([X,Y]);
```

Plot a *phase portrait* (Figure III.3.1 (b)):

```
> DEplot([ds],[x(t),y(t)],t=-10..10,x=-4..4,y=-4..4,
    {[x(0)=0,y(0)=-3.5],[x(0)=4,y(0)=0.2],[x(0)=2,y(0)=0],
    [x(0)=0,y(0)=2]},linecolor=[red,blue,green,brown],
    arrows=small,stepsize=0.02);
```

3.3 Working Themes

Theme 3.1. Solve with Maple the systems in Seminar II.3. Plot the phase portrait and obtain a fundamental matrix for each of those systems.

Theme 3.2. Solve with Maple and plot the phase portrait of the system

$$\begin{cases} x' = ax - y \\ y' = x - ay \end{cases}$$

for several values of parameter a. Comment on the stability of the system.

Theme 3.3. Solve with Maple and find a fundamental matrix of the system

$$\begin{cases} x' = x + z \\ y' = y \\ z' = x - y + z. \end{cases}$$

Draw time series the solution of the system that satisfies the initial conditions

$$x(0) = -1, \, y(0) = 1, \, z(0) = -1,$$

and using the command DEplot3d, plot the three-dimensional phase portrait of the system.

Theme 3.4. Follow the program of Theme 3.3 for the three-dimensional homogeneous linear system having the coefficient matrix

$$\text{(a) } A = \begin{bmatrix} 3 & 2 & 4 \\ 2 & 0 & 2 \\ 4 & 3 & 2 \end{bmatrix}; \quad \text{(b) } A = \begin{bmatrix} 4 & 6 & 6 \\ -3 & -3 & -5 \\ 3 & 1 & 3 \end{bmatrix}.$$

Lab 4 Second-Order Differential Equations

4.1 Spring-Mass Oscillator Equation with Maple

We consider the spring-mass oscillator equation already presented in Section I.3.1:
$$mx'' + cx' + kx = 0.$$
Recall that the term $-cx'$ is the friction force, while $-kx$ is the elastic force of the spring.

We denote by *sol* the solution of the equation subjected to the initial condition *ic*; this has the form '$x(t) =$ Expression'. Making use of the command rhs (right-hand side), the right-hand side of the solution, i.e. the *Expression*, is extracted and named *solE*. Finally, using the command **unapply** we define the function, called *solE*, of variables t, m, c and k, associated with the expression *solE*. This strategy allows us to generate solutions of the equation for different values of the parameters m, c and k. We consider two cases:
(a) Free undamped motion, when $c = 0$ (Figure III.4.1);
(b) Damped oscillations, when $c > 0$ (Figure III.4.2).

In both cases we shall observe the dependence of the mass (undamped or damped) oscillations around the equilibrium $x = 0$ on the elastic force, that is, on the size of parameter k.

```
> with(DEtools):
> with(plots):
> eq:=m*diff(x(t),t$2)+c*diff(x(t),t)+k*x(t)=0;
```

$$eq := m\left(\tfrac{d^2}{dt^2}x(t)\right) + c\left(\tfrac{d}{dt}x(t)\right) + kx(t) = 0$$

```
> ic:=x(0)=1,D(x)(0)=0;
```

$$ic := x(0) = 1, D(x)(0) = 0$$

```
> sol:=dsolve({eq,ic},x(t));
```

$$sol := x(t) = \frac{1}{2}\frac{\left(c\sqrt{c^2 - 4km} + c^2 - 4km\right)e^{\frac{1}{2}\frac{\left(-c+\sqrt{c^2-4km}\right)t}{m}}}{c^2 - 4km}$$
$$+ \frac{1}{2}\frac{\left(c^2 - c\sqrt{c^2 - 4km} - 4km\right)e^{-\frac{1}{2}\frac{\left(c+\sqrt{c^2-4km}\right)t}{m}}}{c^2 - 4km}$$

```
> solE:=rhs(dsolve({eq,ic},x(t)));
> solF:=unapply(solE,t,m,c,k);
```

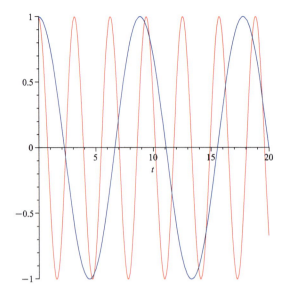

Fig. III.4.1: Undamped oscillations.

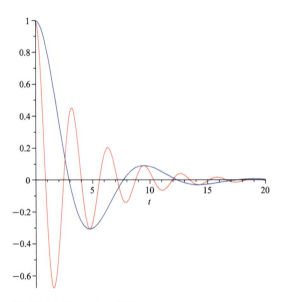

Fig. III.4.2: Damped oscillations.

```
> plot([solF(t,1,0,4),solF(t,1,0,0.5)],t=0..20,
    color=[red,blue]);
> plot([solF(t,1,0.5,4),solF(t,1,0.5,0.5)],t=0..20,
    color=[red,blue]);
```

4.2 Boundary Value Problems with Maple

In Section I.3.6 we considered the boundary value problem

$$\begin{cases} x'' + ax' + bx = h(t), & t \in [0, 1] \\ a_1 x(0) + a_2 x'(0) = 0 \\ b_1 x(1) + b_2 x'(1) = 0, \end{cases}$$

and we found the solution in terms of Green's function. Note that the method based on Green's function also applies to nonconstant coefficient equations. For example, consider the problem

$$\begin{cases} x'' - tx' + 2x = t, & t \in [0, 1] \\ x(0) + x'(0) = 0 \\ -3x(1) + x'(1) = 0, \end{cases}$$

whose solution is $x(t) = t^2 + t - 1$. To solve the problem with Maple and plot the solution we need the following commands:

```
> with(DEtools):
> with(plots):
> eq:=diff(x(t),t$2)-t*diff(x(t),t)+2*x(t)=t;
> bc1:=x(0)+D(x)(0)=0;
> bc2:=-3*x(1)+D(x)(1)=0;
> sol:=dsolve({eq,bc1,bc2}, x(t));
> solE:=rhs(sol);
> solF:=unapply(solE,t);
> plot(solF(t),t=0..1);
```

We note that problems subject to more general boundary conditions can be solved with Maple using the same commands.

4.3 Working Themes

Theme 4.1. Solve with Maple the constant coefficient equations from the examples given in Section I.3.5. In each case, plot one or several solutions associated with different initial conditions.

Theme 4.2. Solve with Maple the constant coefficient equations and the Euler equations considered in Seminar II.4.

Theme 4.3. Continue to analyze the spring-mass oscillations if an additional exterior time-dependent force $f(t)$ acts on the mass and the equation of motion is

$$mx'' + cx' + kx = f(t).$$

Consider some concrete examples for a function f.

Theme 4.4. Solve with Maple the higher-order differential equations:
(a) $x^{(IV)} + x'' = t + 1$;
(b) $x''' - x'' - 8x = e^t - 3$;
(c) $x''' - x'' - 4x' - 4x = 0$, $x(0) = -2$, $x'(0) = 1$, $x''(0) = -5$.

Theme 4.5. Solve with Maple and plot the solution of the following boundary value problems:

(a) $\begin{cases} x'' + 4x' - 5x + t = 0, & t \in [0, 1] \\ x(0) = x(1) = 0 \, ; \end{cases}$

(b) $\begin{cases} x'' + 81x = 1, & t \in [0, 1] \\ x(0) + x'(0) = 1 \\ 3x(1) - 4x'(1) = 5 \, ; \end{cases}$

(c) $\begin{cases} 2tx'' - tx' = 2x = 2t - 6, & t \in [1, 2] \\ x'(1) = 0, \quad x'(2) - x(1) = 6 \, . \end{cases}$

Lab 5 Nonlinear Differential Systems

5.1 The Lotka–Volterra System

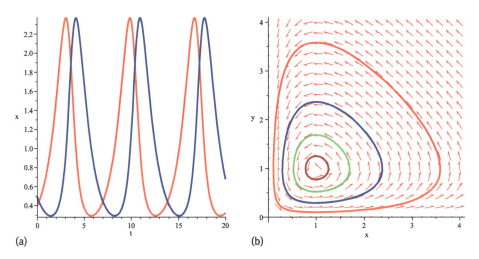

(a) (b)

Fig. III.5.1: The Lotka–Volterra system, (a) time series representation of a solution; (b) phase portrait.

We solve with Maple the Lotka–Volterra system considered in Section I.4.1 for the parameter values $a = b = r = m = 1$. We represent as a 'time series, the solution of an initial value problem (Figure III.5.1 (a)) and we plot the phase portrait of the system (Figure III.5.1 (b)) observing the periodicity of the solutions.

```
> with(DEtools):
> with(plots):
> eq1:=diff(x(t),t)=x(t)-x(t)*y(t);
> eq2:=diff(y(t),t)=-y(t)+x(t)*y(t);
> ds:=eq1,eq2;
> ic:=x(0)=0.4,y(0)=0.5;
> X:=DEplot([ds],[x(t),y(t)],t=0..20,[[ic]],stepsize=0.1,
    linecolor=red,scene=[t,x]):
> Y:=DEplot([ds],[x(t),y(t)],t=0..20,[[ic]],stepsize=0.1,
    linecolor=blue,scene=[t,y]):
> display([X,Y]);
> DEplot([ds],[x(t),y(t)],t=0..10,x=0..4,y=0..4,
    {[x(0)=0.4,y(0)=0.5],[x(0)=0.6,y(0)=0.7],
    [x(0)=0.8,y(0)=0.9],[x(0)=0.2,y(0)=0.3]},arrows=small,
    linecolor=[red,blue,green,brown],stepsize=0.02);
```

5.2 A Model from Hematology

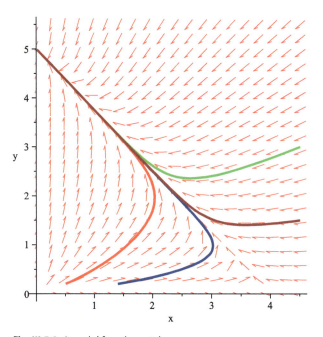

Fig. III.5.2: A model from hematology.

Consider the differential system

$$\begin{cases} x' = \frac{ax}{1+b(x+y)} - cx \\ y' = \frac{Ay}{1+B(x+y)} - Cy, \end{cases} \quad (5.1)$$

which describes the dynamics of two cell populations from bone marrow: normal cells and leukemic cells. The system has three stationary solutions (equilibria): $(0, 0)$, $(d, 0)$ and $(0, D)$, where $d = b^{-1}(ac^{-1} - 1)$, $D = B^{-1}(AC^{-1} - 1)$. Using numerical simulations with Maple, for $b = B = 0.5$, $c = C = 1$ and different values of the parameters a and A, we observe that the normal equilibrium state $(d, 0)$ characterized by the elimination of possible malignant cells is asymptotically stable if $d > D$. In contrast, if $d < D$, then the only stable equilibrium is $(0, D)$, which implies the total elimination of the normal cells.

Simulate the system for each of the following parameter sets: (a) $a = 3$, $A = 3.5$ (when $d = 4$, $D = 5$) (Figure III.5.2; (b) $a = 3.5$, $A = 3$ (when $d = 5$, $D = 4$).

```
> restart:
> with(DEtools):
> with(plots):
```

```
> eq1:=diff(x(t),t)=a*x(t)/(1+0.5*(x(t)+y(t)))-x(t);
> eq2:=diff(y(t),t)=A*y(t)/(1+0.5*(x(t)+y(t)))-y(t);
> ds:=eq1,eq2;
> a:=3:
> A:=3.5:
> DEplot([ds],[x(t),y(t)],t=0..50,x=0..4.5,y=0..5.5,
   {[x(0)=4.5,y(0)=3],[x(0)=4.5,y(0)=1.5],[x(0)=0.5,y(0)=0.2],
   [x(0)=1.4,y(0)=0.2]},linecolor=[red,blue,green,brown],
   arrows=small,stepsize=0.1);
```

5.3 Working Themes

Theme 5.1. Simulate with Maple the Lotka–Volterra predator-prey model including the harvesting of the two species at the same rate E:

$$\begin{cases} x' = x(1-x) - Ex \\ y' = -y(1-x) - Ey \, . \end{cases}$$

Consider several values of E around 1. Explain how the equilibrium changes when parameter E increases. Which species is relatively favored?

Theme 5.2. Analyze with Maple the SIR epidemiologic model (Section I.4.1)

$$\begin{cases} x' = -xy \\ y' = xy - ry \end{cases}$$

subject to the initial condition $x(0) = 0.1$, $y(0) = 0.001$. Consider the following values of the recovery rate r: 0.01; 0.05; 0.07; 0.09. In each case, plot time series of the solution and give interpretation in epidemiological terms.

Theme 5.3. Plot the phase portrait of each of the following nonlinear differential systems:

(a) $\begin{cases} x' = y \\ y' = -x + x^3 \, ; \end{cases}$
(b) $\begin{cases} x' = -y + x(1 - x^2 - y^2) \\ y' = x + y(1 - x^2 - y^2) \, ; \end{cases}$

(c) $\begin{cases} x' = x - xy \\ y' = y - xy \, ; \end{cases}$
(d) $\begin{cases} x' = \frac{1}{2}x + y - \frac{1}{2}(x^3 + xy^2) \\ y' = -x + \frac{1}{2}y - \frac{1}{2}(y^3 + x^2 y) \, . \end{cases}$

Theme 5.4. Realize the phase portraits of the systems considered in Seminar II.7 and graphically explain the theoretical conclusion about the stability of their stationary solutions.

Lab 6 Numerical Computation of Solutions

Most differential equations cannot be solved analytically by simple formulas. Therefore, it is of interest to develop methods of approximation and numerical computation of solutions. Maple's library contains a large number of such methods already implemented and easily accessed by automatic commands. It also allow users to experiment and produce their own procedures and then to import them into Maple. In addition to many available Maple application manuals, the user can find more information, commands and examples in the help menu of the program itself[1].

We present the Maple syntax for numerically solving initial value problems (IVP) and boundary value problems (BVP).

6.1 Initial Value Problems

Example 1. If you try to obtain the solution of the IVP

$$\begin{cases} x'' = \sin(t \sin(x)) \\ x(0) = 1, \quad x'(0) = 0 \end{cases}$$

using the command dsolve as in Lab 2 and Lab 4, then you will get nothing and you are obliged to look for a numerical solution. The commands for computing and plotting the numerical solution on the interval [0, 2] are as follows:

```
> with(DEtools):
> with(plots):
> eq:=diff(x(t),t$2)=sin(t*sin(x(t)));
> ic:=x(0)=1,D(x)(0)=0;
> nsol:=dsolve({eq,ic},x(t),numeric):
> odeplot(nsol,t=0..2);
```

If you desire the value of the numerical solution at some point (say 1.5) use:

```
> nsol(1.5);
```

If you need the values of the solution at a number of points, then use the option output=Array and give a list of points:

```
> nsol:=dsolve({eq,ic},x(t),numeric, output=Array([0,0.5,1,1.5,2]));
```

---

[1] http://www.maplesoft.com/support/help/Maple/view.aspx?path=dsolve/numeric

https://doi.org/10.1515/9783110447446-019

Example 2. Solve numerically the initial value problem

$$\begin{cases} x' = \sin(x+y) \\ y' = \cos(x+y) \\ x(0) = y(0) = 0 \, . \end{cases}$$

Find the solution at $t = 1.2$ and a table of values at the points $0, 0.5, 1, 1.5, 2, 2.5, 3$.

```
> with(DEtools):
> IVP:={diff(x(t),t)=sin(x(t)+y(t)),diff(y(t),t)=cos(x(t)+y(t)),
x(0)=0,y(0)=0};
> nsol:=dsolve(IVP,numeric):
> nsol(1.2);
> nsol:=dsolve(IVP,numeric,output=Array([0,0.5,1,1.5,2,2.5,3]));
```

6.2 Boundary Value Problems

Example 3. Solve numerically the boundary value problem

$$\begin{cases} x'' = \sin(tx), & t \in [0, 3] \\ x(0) = 1, & x(3) = \tfrac{2}{3} \, . \end{cases}$$

Find the value of the numerical solution at $t = 0.5$, plot the solution and obtain a table of its values at the points $0.5, 1, 1.5, 2, 2.5$.

```
> with(DEtools):
> with(plots):
> eq:=diff(x(t),t$2)=sin(t*x(t));
> bc:=x(0)=1,x(3)=2/3;
> nsol:=dsolve({eq,bc},numeric):
> nsol(0.5);
> odeplot(nsol);
> nsol:=dsolve({eq,bc},numeric,output=Array([0.5,1,1.5,2,2.5]));
```

Example 4 (Solving with an initial guess of the solution). For many boundary value problems the solution is not unique and there is no method to obtain all the solutions. However, using different functions as initial guesses we can expect to obtain multiple solutions. For example, consider the boundary value problem

$$\begin{cases} x' = \frac{5}{1+e^{-8(x-1)}} - \tfrac{1}{2}e^{\frac{t}{2}}x, & t \in [0, 1] \\ x(0) = x(1) \, . \end{cases}$$

Take as an initial guess the function $1+t(1-t)$, which satisfies the boundary condition. A numerical solution is obtained using the option approxsoln as follows:

```
> with(DEtools):
> with(plots):
> eq:=diff(x(t),t)=5/(1+exp(-8*(x(t)-1)))-1/2*exp(t/2)*x(t);
> bc:=x(0)=x(1);
> sol1:=dsolve({eq,bc},numeric,approxsoln=[x(t)=1+t*(1-t)]):
> odeplot(sol1);
```

Two other solutions are obtained using as an initial guess the functions $5(1 + t(1 - t))$ and $10^{-1}(1 + t(1 - t))$, respectively. To plot them we add the commands:

```
> sol2:=dsolve({eq,bc},numeric,approxsoln=[x(t)=5*(1+t*(1-t))]):
> odeplot(sol2);
> sol3:=dsolve({eq,bc},numeric,approxsoln=[x(t)=0.1*(1+t*(1-t))]):
> odeplot(sol3);
```

Example 5 (The pendulum equation). Consider the boundary value problem

$$\begin{cases} x'' + a \sin x = f(t), & t \in [0, 1] \\ x(0) = 0, \quad x(1) = b. \end{cases}$$

Recall that the equation describes small oscillations of the pendulum, when $a = g/l$, g is the gravitational constant and l is the length of pendulum (Section I.3.1).
(a) Prove that for $|a| < 4$, the problem has at most one solution.
(b) Plot the numerical solution for $f(t) \equiv 0$, $a = 3$ and $b = \pi/10$. Find the value of the solution in the middle of the interval $[0, 1]$.
(c) For $f(t) \equiv 0$, $a = 30$ and $b = \pi/10$ find three numerical solutions using the option approxsoln with first guesses -1, 0 and 1, respectively. Give a physical interpretation.

Solution. (a) Substituting $y = x - tb$ yields the problem

$$\begin{cases} y'' + a \sin(y + tb) = f(t), & t \in [0, 1] \\ y(0) = 0, \quad y(1) = 0. \end{cases}$$

Then (Section I.3.6)

$$y(t) = a \int_0^1 G(t, s)\, (\sin(y(s) + sb) - f(s))\, ds,$$

where G is the Green function, $G(t, s) = t(1 - s)$ for $0 \le t \le s \le 1$, and $G(t, s) = s(1 - t)$ for $0 \le s < t \le 1$. It is easy to see that

$$0 \le G(t, s) \le s(1 - s) \le \frac{1}{4}.$$

for all $t, s \in [0, 1]$. If y_1 and y_2 are two solutions, then

$$|y_1(t) - y_2(t)| \le |a| \int_0^1 G(t,s) |\sin(y_1(s) + sb) - \sin(y_2(s) + sb)|\, ds$$

$$\le \frac{1}{4} |a| \max_{s \in [0,1]} |y_1(s) - y_2(s)|\ .$$

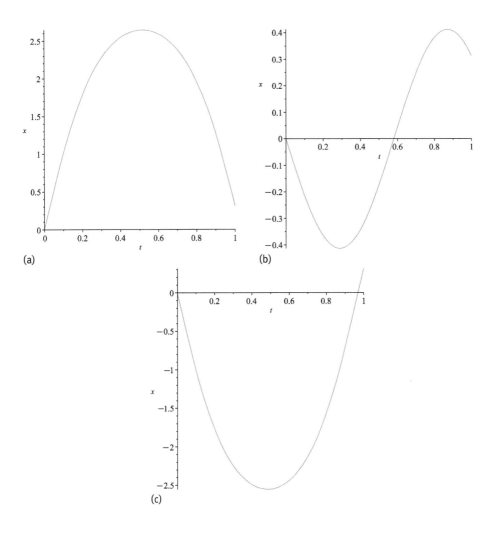

Fig. III.6.1: Three solutions to a BVP for the pendulum equation, (a) solution 1; (b) solution 2; (c) solution 3.

Taking the maximum for $t \in [0, 1]$ yields

$$\max_{t\in[0,1]} |y_1(t) - y_2(t)| \le \frac{1}{4} |a| \max_{t\in[0,1]} |y_1(t) - y_2(t)|.$$

This, in view of $|a|/4 < 1$, implies $\max_{t\in[0,1]} |y_1(t) - y_2(t)| = 0$, hence $y_1 = y_2$.

(b) Maple code:

```
> with(DEtools):
> with(plots):
> eq:=diff(x(t),t$2)+a*sin(x(t))=0;
> bc:=x(0)=0,x(1)=b;
> a:=3;
> b:=Pi/10;
> sol:=dsolve({eq,bc},numeric,approxsoln=[x(t)=1]):
> odeplot(sol);
> sol(1/2);
```

(c) The graphs of the three solutions are presented in Figure III.6.1.

6.3 Working Themes

Theme 6.1. Solve numerically the pendulum equation $x'' + \sin x = 0$ subject to
(a) the initial conditions $x(0) = 1$, $x'(0) = 0$;
(b) the boundary conditions $x(0) = 1$, $x'(1) = 0$.
Repeat the analysis for the pendulum motion under an exterior force $-1.5e^{-t}$, that is for the equation $x'' + \sin x - 1.5e^{-t} = 0$. Compare and comment on the two results.

Theme 6.2. Consider the boundary value problem

$$\begin{cases} x'' = a \sin x + by \\ y'' = cx + d \cos y \quad (t \in [0, 1]) \\ x(0) = x(1) = y(0) = y(1) = 0. \end{cases}$$

(a) Prove that the problem has at most one solution if for the matrix

$$M = \frac{1}{4} \begin{bmatrix} a & b \\ c & d \end{bmatrix},$$

M^k tends to the zero matrix as $k \to +\infty$.

(b) Solve with Maple and determine $x(0.5)$ and $y(0.5)$ for $a = 1$, $b = 5$, $c = -1/4$ and $d = -1$.

Hint. (a) If (x, y) and $(\bar{x}, \bar{y})$ are two solutions, then using Green's function G and the inequality $G \leq 1/4$ (see Example 5 above) one finds that

$$\|x - \bar{x}\| \leq \frac{1}{4}(a\|x - \bar{x}\| + b\|y - \bar{y}\|),$$

$$\|y - \bar{y}\| \leq \frac{1}{4}(c\|x - \bar{x}\| + d\|y - \bar{y}\|),$$

where $\|u\| = \max_{t \in [0,1]} |u(t)|$. These inequalities can be put under the vector form

$$\begin{bmatrix} \|x - \bar{x}\| \\ \|y - \bar{y}\| \end{bmatrix} \leq M \begin{bmatrix} \|x - \bar{x}\| \\ \|y - \bar{y}\| \end{bmatrix},$$

which applied successively yields

$$\begin{bmatrix} \|x - \bar{x}\| \\ \|y - \bar{y}\| \end{bmatrix} \leq M^k \begin{bmatrix} \|x - \bar{x}\| \\ \|y - \bar{y}\| \end{bmatrix}$$

for all k. Letting $k \to +\infty$ gives $\|x - \bar{x}\| = \|y - \bar{y}\| = 0$. Hence the two solutions are equal.

Lab 7 Writing Custom Maple Programs

As demonstrated previously, we are able to automatically solve a given problem and get its solution by using Maple with simple commands. Behind these commands there are numerous hidden procedures, already implemented and continuously updated, which are stocked and organized into packages. In general, we are glad to get the result and we do not think about the theory behind it.

However, the reader who likes to experiment or use a particular algorithm is free to write down a custom program and use it with Maple. We shall present a program for Maple implementation of the method of successive approximations, Euler's method for numerical solving of initial value problems, and the shooting method for boundary value problems.

7.1 Method of Successive Approximations

Example 1 (One equation). We present the Maple program for the successive approximations of the solution to the initial value problem

$$x' = -2x + 2t^2 + 1, \quad x(0) = 1.$$

For comparison, we shall visualize (Figure III.7.1) the graphs of the exact solution $x = t^2 - t + 1$ and of its approximations x_1, x_2 and x_3, on the interval [0, 2/5]. Notice that for this problem, on each interval $[a, b]$, the conditions from Section I.4.8.3 hold and this guarantees the convergence of the sequence of successive approximations to the exact solution.

At each step, the method involves the calculation of a definite integral. This generally requires an additional approximation procedure for definite integrals, which makes the method less useful for applications. In order that the expression of the definite integral be effectively computed and put in a simpler form, we must use the command `simplify`. For a loop (multiple executions of one or more commands) we use the `for` statement, whose syntax is

for counter from start to end by step do instructions od;

The statement 'by step' can be omitted if the step is 1.

```
> with(DEtools): with(plots):
> f:=(t,x)→ -2*x+2*t**2+1;
> x[0]:=t→1;
> n:=6;
```

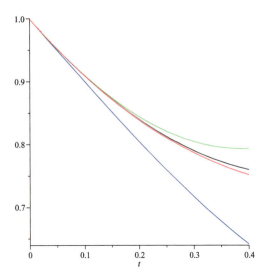

Fig. III.7.1: Picard iterations.

```
> for k from 1 to n do x[k]:=unapply(simplify(1+int(f(s,x[k-1](s)),
  s=0..t)),t) od;
```

$$x_1 := t \to 1 - t + \frac{2}{3}t^3$$

$$x_2 := t \to 1 - t + t^2 - \frac{1}{3}t^4 + \frac{2}{3}t^3$$

$$x_3 := t \to 1 - t + t^2 + \frac{2}{15}t^5 - \frac{1}{3}t^4$$

$$x_4 := t \to 1 - t + t^2 - \frac{2}{45}t^6 + \frac{2}{15}t^5$$

$$x_5 := t \to 1 - t + t^2 + \frac{4}{315}t^7 - \frac{2}{45}t^6$$

$$x_6 := t \to 1 - t + t^2 - \frac{1}{315}t^8 + \frac{4}{315}t^7$$

```
> eq:=diff(y(t),t)=-2*y(t)+2*t**2+1;
```

$$ec := \tfrac{d}{dt}y(t) = -2y(t) + 2t^2 + 1$$

```
> ic:=y(0)=1:
> dsolve({eq,ic},y(t));
```

$$y(t) = 1 - t + t^2$$

```
> plot([t**2-t+1,x[1](t),x[2](t),x[3](t)],t=0..2/5,color=
  [black,blue,green,red]);
```

Example 2 (Systems of equations). Generate successive approximations of the solution to the initial value problem

$$\begin{cases} x' = x - y + t \\ y' = x + y - 1 \\ x(0) = 1, \quad y(0) = 2 \end{cases}$$

using the commands:

```
> with(DEtools):
> f:=(t,x,y)→x-y+t;
> g:=(t,x,y)→x+y-1;
> x[0]:=t→1;
> y[0]:=t→2;
> n:=3;
> for k from 1 to n do
    x[k]:=unapply(simplify(1+int(f(s,x[k-1](s),y[k-1](s)),s=0..t)),t):
    y[k]:=unapply(simplify(2+int(g(s,x[k-1](s),y[k-1](s)),s=0..t)),t):
  od;
```

$$x_1 := t \to 1 - t + \frac{1}{2}t^2$$

$$y_1 := t \to 2 + 2t$$

$$x_2 := t \to 1 - t - t^2 + \frac{1}{6}t^3$$

$$y_2 := t \to 2 + 2t + \frac{1}{2}t^2 + \frac{1}{6}t^3$$

$$x_3 := t \to 1 - t - t^2 - \frac{1}{2}t^3$$

$$y_3 := t \to 2 + 2t + \frac{1}{2}t^2 - \frac{1}{6}t^3 + \frac{1}{12}t^4$$

7.2 Euler's Method

Euler's method is a numerical tool for approximating values for solutions of differential equations (see, e.g. [5, p. 31], [15, pp. 74, 229], [26, p. 106]). The continuous model expressed by a differential equation is replaced by an approximate discrete model, which is easy to implement on a computer.

The idea of Euler's method for numerical approximation of the solution of the Cauchy problem

$$x' = f(t, x) \quad (t_0 \le t \le T), \quad x(t_0) = x_0$$

is to discretize the interval $[t_0, T]$ by dividing it into n equal parts of length $h = (T - t_0)/n$, and to determine an approximation x_k of the exact solution at each point of division $t_k = t_0 + kh$, $k = 0, 1, \ldots, n$. If $x(t_k)$ is the value of the exact solution at t_k, then the value $x(t_{k+1})$ is given by

$$x(t_{k+1}) = x(t_k) + \int_{t_k}^{t_{k+1}} f(s, x(s))\, ds \,.$$

Approximating the integral by $(t_{k+1} - t_k)f(t_k, x(t_k))$ yields the approximation formula

$$x(t_{k+1}) \simeq x(t_k) + hf(t_k, x(t_k)) \,.$$

This suggests the iterative procedure called *Euler's method* for computing the approximations x_k:

$$x_{k+1} = x_k + hf(t_k, x_k),$$

$k = 0, 1, \ldots, n - 1$. Here, the starting value x_0 is given by the initial condition.
As an example, consider the initial value problem

$$x' = x^2 - t^3 \ (1 \le t \le 4), \quad x(1) = \frac{3}{2}.$$

The Maple program to generate approximations of the exact solution at 11 equidistant points of the interval $[1, 4]$ is:

```
> f:=(t,x)→x**2-t**3:
> t0:=1: T:=4: x0:=3/2:
> n:=10: h:=evalf((T-t0)/n):
> t:=t0: x:=x0:
> for k from 1 to n do x:=x+h*f(t,x): t:=t+h: print(t,x): od;
```

The output below contains the points t_k in the left column, and the approximants x_k in the right column:

1.300000000	1.875000000
1.600000000	2.270587500
1.900000000	2.588457778
2.200000000	2.540791878
2.500000000	1.283078888
2.800000000	−2.910533683
3.100000000	−6.954771787
3.400000000	−1.381416604
3.700000000	−12.60012305
4.000000000	19.83290722

7.3 The Shooting Method

The *shooting method* is a technique for solving a boundary value problem by reducing it to an initial value problem. The idea is to try ('shoot') initial values until the solution of the corresponding Cauchy problem satisfies the desired boundary condition.

Assume that we are interested in one solution of a first-order differential system

$$x' = f(t, x), \quad t \in [t_0, t_1], \tag{7.1}$$

which satisfies the *multipoint boundary condition*

$$g(x(s_1), x(s_2), \ldots, x(s_m)) = 0, \tag{7.2}$$

where $f: [t_0, t_1] \times \mathbb{R}^n \to \mathbb{R}^n$, $x = [x_1, x_2, \ldots, x_n]^T$, $g: \mathbb{R}^{nm} \to \mathbb{R}^p$ and $t_0 \leq s_1 < s_2 < \ldots < s_m \leq t_1$ are fixed points.

The idea is to solve the initial value problem

$$\begin{cases} x' = f(t, x), & t \in [x_0, x_1] \\ x(x_0) = v, \end{cases} \tag{7.3}$$

for some initial value $v \in \mathbb{R}^n$. Assume that the Cauchy problem (7.3) has for each v a unique solution $x(., v)$. Then we may define $G: \mathbb{R}^n \to \mathbb{R}^p$,

$$G(v) = g(x(s_1, v), x(s_2, v), \ldots, x(s_m, v)).$$

Obviously, the function $x(., v)$ solves the multipoint boundary value problem (7.1)–(7.2) if v is a root of G, that is

$$G(v) = 0. \tag{7.4}$$

Thus the problem is reduced to solving the nonlinear algebraic equation (7.4). To this purpose one may employ any of the numerical methods for finding roots, such as Newton's method. In particular, if $n = p = 1$, one may use the secant method or the bisection method. The chosen method for finding roots, basically an iterative procedure, will dictate the rules of the 'game'. Thus v is changed from one value v_k to another value v_{k+1} as prescribed by the algorithm, and the 'game' stops when the deviation of $G(v_k)$ from zero becomes less than an accepted tolerance.

Note that the method also applies to boundary value problems for higher-order differential equations. In such cases, the equations are automatically transformed into first-order systems.

Maple implementation is simple if the solution of the initial value problem has an explicit exact formula, as in the example below.

Example. Apply the shooting method to solve the boundary value problem

$$\begin{cases} x'' + x = t^2 - t + 2 \\ x(0) = 0, \quad x(2) = 2. \end{cases}$$

7.3 The Shooting Method

```
> with(DEtools):
> eq:=diff(x(t),t$2)+x(t)=t**2-t+2;
> ic:=x(0)=0, D(x)(0)=v;
```

Solve the initial value problem:

```
> dsolve({eq,ic},x(t)));
```

$$x(t) = \sin(t)(v + 1) + t^2 - t$$

Put the boundary condition at $t = 2$, $x(2) = 2$ and solve in v:

```
> fsolve(sin(2)(v+1)+2=2);
```

$$-1.$$

Hence the solution of the BVP is the solution of the IVP with $v = -1$:

```
> finalic:=x(0)=0, D(x)(0)=-1;
> dsolve({eq,finalic},x(t));
```

The obtained solution $t^2 - t$ is the exact solution of the problem.

If the solution of the initial value problem does not have an explicit exact formula, then one can only expect to produce numerical solutions. To this aim, a 'numerical' procedure for the shooting method is necessary.

In general, a Maple procedure starts with

$$\text{name} := \text{proc}(\text{parameters}),$$

continues by commands operating on the parameters, and ends by end proc;

Below is such a procedure[1] for numerically solving second-order differential equations subject to the boundary conditions

$$x(t_0) = a, \quad x(t_1) = b.$$

The procedure works under the assumption that two initial values v_1 and v_2 are known such that the numbers

$$x(t_1, v_1) - b \quad \text{and} \quad x(t_1, v_2) - b$$

have opposite signs. Such values v can be found by trial and error, after providing the equation and the input data a, b, t_0, t_1 and v, by using the commands:

```
> sol0:=dsolve({eq,x(t0)=a,D(x)(t0)=v},numeric):
> sol0(t1); rhs(sol0(t1)[2])-b;
```

Notice that the output sol0 is the triple $[t_1, x(t_1), x'(t_1)]$, and the command rhs(sol0(t1)[2]) is aimed at extracting its second component $x(t_1)$.

---

[1] Written by Marcel-Adrian Șerban

As an example, consider the problem

$$\begin{cases} x'' + 30\sin x = 0, & t \in [0, 1] \\ x(0) = 0, & x(1) = \frac{\pi}{10}. \end{cases}$$

```
> shooting_meth:=proc(eq,t0,t1,a,b,v1,v2,err)
    local v,v11,v22, sol,sol1,sol2:
    v11:=v1
    v22:=v2
    sol1:=dsolve({eq,x(t0)=a,D(x)(t0)=v11},numeric)
    sol2:=dsolve({eq,x(t0)=a,D(X)(t0)=v22},numeric)
    v:=evalf((v11+v22)/2)
    sol:=dsolve({eq,x(t0)=a,D(x)(t0)=v},numeric)
    while abs(rhs(sol(t1)[2])-b)>err
       do
       if (rhs(sol(t1)[2])-b)*(rhs(sol1(t1)[2])-b)<0
          then
          v22:=v
          sol2:=dsolve({eq,x(t0)=a,D(x)(t0)=v22},numeric)
             else
                 v11:=v
          sol1:=dsolve({eq,x(t0)=a,D(x)(t0)=v11},numeric)
             end if
             v:=evalf(v11+v22)/2
             sol:=dsolve({eq,x(t0)=a,D(x)(t0)=v},numeric)
       end do
    end;
```

Here is the example:

```
> eq:=diff(x(t),t$2)+30*sin(x(t))=0;
> bc:=x(0)=0,x(1)=Pi/10;
> Digits:=20;
> t0:=0;t1:=1;a:=0;b:=Pi/10;v1:=5;v2:=20;
> sol1:=dsolve({eq,x(t0)=a,D(x)(t0)=v1},numeric):
> sol1(t1); rhs(sol1(t1)[2])-b;
> sol2:=dsolve({eq,x(t0)=a,D(x)(t0)=v2},numeric):
> sol2(t1); rhs(sol2(t1)[2])-b;
> err:=10^(-12);
> sol_shm:=shooting_meth(eq,t0,t1,a,b,v1,v2,err):
> sol_shm(t1);
> with(plots):
> odeplot(sol_shm,[t,x(t)],t0..t1);
```

The result is the numerical solution (a) from Example 5 (c) in Lab 6. To obtain the other two solutions, repeat the commands for $v1 = -5$, $v2 = -1$ and $v1 = -15$, $v2 = -5$.

The above procedure includes the conditional statement `if`, which is important in programming in order to branch the computation according to some condition. The syntax of this statement is

```
if condition(s) then command(s) else command(s) end if.
```

7.4 Working Themes

Theme 7.1. Apply with Maple the method of successive approximations to the Cauchy problems in Exercises 6.1 and 6.4 from Seminar II.6.

Theme 7.2. Use Euler's method to numerically solve the following Cauchy problems:

(a) $\begin{cases} x' = \sqrt[3]{x} - t, & t \in [0, 1] \\ x(0) = 1; \end{cases}$
(b) $\begin{cases} x' = \frac{t}{x^2+1} - x^3, & t \in [0, 2] \\ x(0) = 0, \end{cases}$

and compare with the results given by the successive approximations method.

Theme 7.3. Use Euler's method to numerically solve the initial value problem $x' = x \sin t$ ($0 \le t \le 20$), $x(0) = 1$, with the number of steps $n = 50, 100$ and 200. Compare the numerical solutions to the exact solution.

Theme 7.4 ([15, p. 80]). A population of bacteria, given in millions of organisms, is governed by the law

$$x' = 0.6x \left(1 - \frac{x}{k(t)}\right),$$

where the carrying capacity is time dependent, given by $k(t) = 10 + 0.9 \sin t$, and time is measured in days. Plot the bacteria population for the first 40 days if the initial population was $x(0) = 0.2$. Using Euler's method, find numerical estimates of the bacteria population every four days. Then solve the problem using the automatic Maple procedure based on the option `numeric` and compare the results.

Theme 7.5. Apply the shooting method to the boundary value problem

$$\begin{cases} t^2 x'' - 3tx' + 3x = 0 \\ x'(0) = 1, \quad x(1) = 0. \end{cases}$$

Theme 7.6. Use the Maple procedure for the shooting method to numerically solve the boundary value problem

$$\begin{cases} x'' + \frac{1}{t+1} x^2 \sin\left(\frac{1}{2} x \ln(x + 1)\right) = 0 \\ x(0) = 5, \quad x(1) = 3. \end{cases}$$

As an interval for the location of v, take $[-6, 6]$.

Lab 8 Differential Systems with Control Parameters

8.1 Bifurcations

Example (Visualization to Examples 6.6 and 6.7 from Section I.6.2). Plot the phase portrait of the following differential systems for $p < 0$ and $p > 0$, and comment on the bifurcation at $p = 0$:

$$(1) \begin{cases} x' = -y + x(p - x^2 - y^2) \\ y' = x + y(p - x^2 - y^2) \, ; \end{cases} \quad (2) \begin{cases} x' = -y - x(p - x^2 - y^2) \\ y' = x - y(p - x^2 - y^2) \, . \end{cases}$$

For the first system and $p = -4$, use the below commands, then change the value of p to $+4$ (Figure III.8.1). For the second system repeat the simulations for $t = -100..0$ (Figure III.8.2).

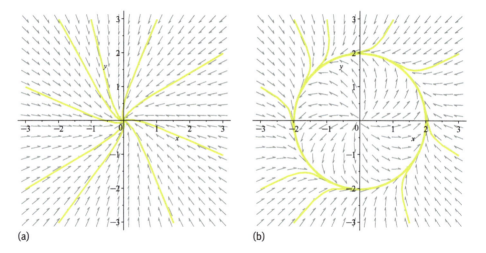

(a)   (b)

Fig. III.8.1: Phase portrait of system (1): (a) $p = -4$, (b) $p = +4$.

```
> with(DEtools): with(plots):
> p:=-4;
> eq1:=diff(x(t),t)=-y(t)+x(t)*(p-x(t)**2-y(t)**2);
> eq2:=diff(y(t),t)=x(t)+y(t)*(p-x(t)**2-y(t)**2);
> DEplot([eq1,eq2],[x,y],t=0..100,x=-3..3,y=-3..3,
{[x(0)=-3,y(0)=1], [x(0)=1,y(0)=3], [x(0)=3,y(0)=2], [x(0)=-1, y(0)=3],
[x(0)=-2,y(0)=3],[x(0)=-2,y(0)=-3], [x(0)=-3,y(0)=-2],
[x(0)=3,y(0)=-1]},stepsize=0.05);
```

https://doi.org/10.1515/9783110447446-021

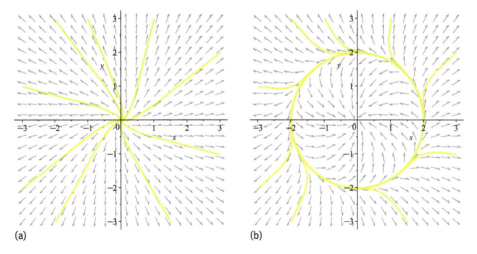

Fig. III.8.2: Phase portrait of system (2): (a) $p = -4$, (b) $p = +4$.

8.2 Optimization with Maple

The aim of this lab course is to solve with Maple the optimization problems presented in Section I.6.3. These problems are set at the border between differential equation theory and optimization theory. For this type of problems there is a specific package that must be loaded before any calculation, namely Optimization[1].

Example 1 (Maple solution to Example I.6.8). Consider the one-species harvesting model
$$x' = ax\left(1 - \frac{x}{k}\right) - ux$$
with control parameters k (carrying capacity) and u (harvesting effort). To solve with Maple the optimization problem (6.4) for given numerical values to parameters a, c, x_L, u_L and k_L, we need the commands

```
> with(Optimization):
> a:=10;
> c:=300;
> xL:=10;
> uL:=0.02;
> kL:=10**5;
> Maximize(c*u/k, {(a-u)*k>=a*xL, u>=uL, k<=kL},
assume=nonnegative);
```

---

[1] http://www.maplesoft.com/support/help/Maple/view.aspx?path=examples/Optimization;
P. E. Fishback, Linear and Nonlinear Programming with Maple: An Interactive, Applications-Based Approach, CRC Press, 2009.

which yield the output

$$[75,0000000000000711, [k = 20.0000002552859, u = 5.00000006382149]] .$$

Example 2 (Maple solution to Example I.6.9). The Lotka–Volterra model with harvesting is

$$\begin{cases} x' = x(r - ay) - u_1 x \\ y' = -y(m - bx) - u_2 y . \end{cases}$$

The control parameters are u_1 and u_2 and the optimization problem considered in Section I.6.3 is to maximize the profit

$$\varphi = c_1 u_1 \frac{m + u_2}{b} + c_2 u_2 \frac{r - u_1}{a}$$

under the constraints

$$\alpha \le \frac{m + u_2}{b} \cdot \frac{a}{r - u_1} \le \beta ,$$

where α and β are two given numbers with $\alpha < ma/(br) < \beta$.

Take the parameter values $r = 2$, $m = 1$, $a = 1$, $b = 0.1$, $c_1 = 100$, $c_2 = 150$, $\alpha = 5$ and $\beta = 15$. Then, the commands

```
> with(Optimization):
> Maximize(1000*(1+u2)*u1+150*(2-u1)*u2, {0.5*(2-u1)<=1+u2, 1+u2
<=1.5*(2-u1)}, assume=nonnegative);
```

give the result

$$[1592.647\ldots, [u1 = 0.882\ldots, u2 = 0.676\ldots]] .$$

Example 3 (Maple solution to Example I.6.10). Consider the differential system modeling normal-leukemic cell dynamics

$$\begin{cases} x' = \frac{ax}{1+b(x+y)} - cx \\ y' = \frac{u_1 Ay}{1+u_2 B(x+y)} - u_3 Cy \end{cases}$$

with control parameters u_1, u_2 and u_3. The problem is to minimize the objective function

$$\varphi = c_1(1 - u_1) + c_2(u_2 - 1) + c_3(u_3 - 1)$$

subject to the constraints

$$\frac{1}{u_2 B}\left(\frac{u_1 A}{u_3 C} - 1\right) \le d = \frac{1}{b}\left(\frac{a}{c} - 1\right), \quad u_1 \le 1, \quad u_2 \ge 1, u_3 \ge 1 .$$

Take the following parameter values: $a = 0.23$, $A = 0.45$, $b = B = 2.2 \times 10^{-8}$, $c = C = 0.01$, when $d = 10^9$, and choose $c_1 = 40$, $c_2 = 20$, $c_3 = 15$. Using the Maple commands

```
> with(Optimization):
> Minimize(40*(1-u1)+20*(u2-1)+15*(u3-1), {u1<=1, u2>=1, u3>=1,
u1*0.45-u3*0.01<=0.22*u2*u3}, assume=nonnegative);
```

the output is

$$[13.634\ldots, [u1 = 1, u2 = 1.193\ldots, u3 = 1.651\ldots]]$$

8.3 Working Themes

Theme 8.1. Consider the following equations:

$$\text{Rayleigh's equation:} \quad x'' + x'^3 - px' + x = 0,$$
$$\text{Van der Pol's equation:} \quad x'' - (p - x^2)x' + x = 0.$$

For each equation plot the phase portrait of the equivalent first-order differential system for $p < 0$ and $p > 0$, and check that there is a Hopf bifurcation at $p = 0$.

Theme 8.2. Consider the SIR epidemic model

$$\begin{cases} S' = -aSI \\ I' = aSI - rI, \end{cases}$$

where at each time t, the numbers $S(t)$, $I(t)$ and $R(t)$ denote the number of individuals of a fixed size community of N members, who can either get the illness, or are infected and have immunity, respectively. Assume that the disease stops spreading in time, that is $I(t) \to 0$ as $t \to +\infty$. Denote by I_0 the initial number of infective individuals and by R_0 the number of initial vaccinated individuals.

(a) Prove that the final size of population affected by the disease $R_\infty - R_0$, where $R_\infty = \lim_{t \to +\infty} R(t)$, can be obtained by solving the equation

$$R_\infty = R_0 + \frac{r}{a} \ln \frac{N - R_0 - I_0}{N - R_\infty}. \tag{8.1}$$

(b) What percentage of the population should be vaccinated at the beginning of the epidemic to be sure that the affected population will not exceed 25% of N?

(c) Assume that the population splits into two classes of fixed sizes N_1 and N_2 ($N = N_1 + N_2$), for instance, the classes of young and old people, and that each class has its own transmission and recovery rates, a_1, r_1 and a_2, r_2, respectively. What percentages of the two classes, expressed by integer numbers p and q should be vaccinated at the beginning of the epidemic such that the total number of vaccinations $(pN_1 + qN_2)/100$ is minimal and the joint affected population will not exceed 25% of N? Write a pseudocode algorithm to give the answer and try to convert it into Maple code.

Hint. (a) From $S(t) + I(t) + R(t) = N$ (constant), one has $R'(t) = -S'(t) - I'(t)$, such that $R' = rI$. The equations of R' and I', by division, give

$$\frac{dR}{dI} = \frac{r}{a(N - R - I) - r}.$$

Substitution $u := R + I$ yields the separable equation

$$\frac{du}{dI} = \frac{a(N - u)}{a(N - u) - r}.$$

Integrating gives

$$u - u_0 + \frac{r}{a} \ln \frac{N - u}{N - u_0} = I - I_0,$$

or

$$R - R_0 + \frac{r}{a} \ln \frac{N - R - I}{N - R_0 - I_0} = 0.$$

Letting $t \to +\infty$ and using $I_\infty = 0$ one obtains (8.1).

(b) The problem is to find R_0 such that $R_\infty - R_0 \leq N/4$. Let us look for R_0 such that $R_\infty - R_0 = N/4$. Replacing R_∞ by $R_0 + N/4$ in (8.1) yields the minimum number of individuals to be vaccinated

$$R_0 = \frac{I_0 + N\left(\frac{3}{4} e^{\frac{aN}{4r}} - 1\right)}{e^{\frac{aN}{4r}} - 1}.$$

This result is valid if $0 < I_0 < N/4$ and the solution R_∞ of (8.1) corresponding to $R_0 = 0$ (no vaccination) is greater than $N/4$.

(c) Let I_{10}, I_{20} and R_{10}, R_{20} be the numbers of infected and vaccinated individuals, respectively, in the two classes at the beginning of epidemic, and let $R_{i\infty} = \lim_{t \to +\infty} R_i(t)$ for $i = 1, 2$. The problem is to find integers p, q with $0 \leq p, q \leq 100$, such that $pN_1 + qN_2$ is minimal and $R_{1\infty} - R_{10} + R_{2\infty} - R_{20} \leq N/4$, that is

$$R_{1\infty} - \frac{pN_1}{100} + R_{2\infty} - \frac{qN_2}{100} \leq \frac{N}{4}.$$

Bibliography

[1] R. P. Agarwal and D. O'Regan, *An Introduction to Ordinary Differential Equations*, Springer, New York, 2008.
[2] V. Barbu, *Differential Equations*, Springer, Berlin, 2016.
[3] D. Betounes, *Differential Equations. Theory and Applications with Maple*, Springer, New York, 2000.
[4] T. A. Burton, *Stability and Periodic Solutions of Ordinary and Functional Differential Equations*, Dover Publications, Mineola and New York, 2005.
[5] S. L. Campbell and R. Haberman, *Introduction to Differential Equations*, Princeton University Press, Princeton and Oxford, 2008.
[6] B. D. Craven, *Control and Optimization*, Chapman & Hall, London, 1995.
[7] J. H. Davis, *Differential Equations with Maple*, Birkhäuser, Boston, 2001.
[8] J. K. Hale, *Ordinary Differential Equations*, Dover, 2009.
[9] M. W. Hirsch, S. Smale and R. L. Devaney, *Differential Equations, Dynamical Systems & An Introduction to Chaos*, Elsevier, New York, 2004.
[10] S.-B. Hsu, *Ordinary Differential Equations with Applications*, World Scientific, New Jersey, 2006.
[11] J. H. Hubbard and B. H. West, *Differential Equations: A Dynamical Systems Approach*, Springer, New York, 1991.
[12] D. S. Jones, M. J. Plank and B. D. Sleeman, *Differential Equations and Mathematical Biology*, CRC Press, 2010.
[13] D. Kaplan and L. Glass, *Understanding Nonlinear Dynamics*, Springer, New York, 1995.
[14] Y. A. Kuznetsov, *Elements of Applied Bifurcation Theory*, Springer, Berlin, 1995.
[15] J. D. Logan, *A First Course in Differential Equations*, Springer, New York, 2006.
[16] H. Logemann and E. P. Ryan, *Ordinary Differential Equations*, Springer, London, 2014.
[17] S. Lynch, *Dynamical Systems with Applications using Maple*, Birkhäuser, Boston, 2010.
[18] L. Perko, *Differential Equations and Dynamical Systems*, Springer, Berlin, 1991.
[19] L. C. Piccinini, G. Stampacchia and G. Vidossich, *Ordinary Differential Equations in $\mathbb{R}^n$*, Springer, New York, 1984.
[20] R. Precup, *Methods in Nonlinear Integral Equations*, Springer Science + Business Media, Dordrecht, 2002.
[21] R. M. Redheffer, *Differential Equations: Theory and Applications*, Jones and Bartlett Publishers, Boston, 1991.
[22] J. C. Robinson, *An Introduction to Ordinary Differential Equations*, Cambridge Univ. Press, Cambridge, 2004.
[23] C. C. Ross, *Differential Equations. An Introduction with Mathematica*, Springer, New York, 2004.
[24] I. A. Rus, *Differential Equations, Integral Equations and Dynamical Systems* (in Romanian), Transilvania Press, Cluj, 1996.
[25] S. H. Strogatz, *Nonlinear Dynamics and Chaos*, Addison–Wesley, Reading, 1994.
[26] I. I. Vrabie, *Differential Equations. An Introduction to Basic Concepts, Results and Applications*, World Scientific, Singapore, 2016.
[27] P. Waltman, *A Second Course in Elementary Differential Equations*, Academic Press, London, 1986.

Index

A
asymptotically stable equilibrium 52
asymptotically stable solution 97
asymptotically stable system 99
autonomous equation 4

B
basin of attraction 111
Bernoulli equation 13
bicompartmental model 24, 146
bifurcation 113
bifurcation diagram 114
bifurcation point 113
Bihari–LaSalle inequality 159
bilocal conditions 66
boundary conditions 66
boundary value problem 66

C
carrying capacity 19
Cauchy problem 4, 22, 69
center 51
center manifold 176
center subspace 174
characteristic equation 36, 59
Clairaut's equation 132
conservative force 56
constant coefficient equation 5
constant coefficient linear system 23
constraint optimization problem 122
contamination model 147
continuous dependence of solutions 81
control parameter 113
critical point 47

D
damped pendulum equation 54
damped spring-mass oscillator equation 54
decay constant 16
differential equation of a family of curves 131
differential equation of orthogonal trajectories 131
differential system, n-dimensional 22
direction field 47
dynamic optimization 122

E
eigenvalue 36
eigenvector 36
energy conservation theorem 56
equilibrium 47
equilibrium criterion 88
Euler equation 64
Euler's method 207

F
fixed point of a system 47
flow 173
focus 51
fundamental matrix 32

G
general form 3
general solution 4
global existence and uniqueness 94
global stable manifold 175
global unstable manifold 175
globally asymptotically stable system 111
glucose-insulin interaction 146
Gompertz equation 20
Green's function 68
Gronwall's inequality 75
growth rate 18
growth rate, per capita 18

H
half-life 16
Harnack type inequality 162
heat conduction 152
heat transfer coefficient 17
hematological model 75
higher-order equation 69
homogeneous equation 5
homogeneous system 23
homogeneous type 7
Hopf bifurcation 115
Hurwitzian matrix 101

I
immunological model 74
implicit solution 8
initial data 22
initial point 22

initial value problem 4
initial values 22
integrating factor 9
invariant set 174

J
Jacobian matrix 104
Jordan block 44
Jordan canonical form 44
Jordan decomposition theorem 44

K
kinetic energy 56

L
Lagrange's equation 132
linear differential equation 5
linear differential system 22
linear equation 8
linearization 25
Lipschitz continuous 78
local existence and uniqueness 93
locally Lipschitz continuous 78
logistic equation 19
Lorenz system 171
Lotka–Volterra model 73
Lyapunov function 107
Lyapunov stability 97
Lyapunov's direct method 106
Lyapunov's linearization method 104

M
Malthus model 18
Maple
– animate 184
– animate3d 184
– approxsoln 200
– arrows=none 185
– DEplot 185
– DEplot3d 190
– dfieldplot 185
– discont=true 183
– display 190
– dsolve 185
– eigenvalues 188
– eigenvectors 188
– evalf 180
– exponential 189
– for-loop 204

– if-conditional 211
– implicitplot 183
– inverse 188
– linsolve 188
– matrix 188
– Maximize 213
– Minimize 214
– output=Array 198
– plot 183
– plot3d 184
– proc 209
– rhs 191
– scaling=constrained 183
– simplify 204
– unapply 191
– with(DEtools) 185
– with(linalg) 188
– with(Optimization) 213
– with(plots) 183
matrix exponential 29
matrix notation 26
maximal domain 3
method of elimination 58
multipoint condition 208

N
negatively invariant set 174
Newton's law 53
Newton's law of heat transfer 17
node 49
nonautonomous equation 4
nonhomogeneous equation 5
nonhomogeneous part 5
nonhomogeneous term 23
normal form 3

O
optimal control 122
optimal one-species harvesting 119, 213
optimal treatment of leukemia 121, 214
optimal two-species harvesting 120, 214
orbit 46
ordinary differential equation 3
orthogonal trajectories 131

P
pendulum equation 54
periodic solution 85
perturbation term 101

perturbed linear system 101
phase plane 46
phase portrait 46
Picard's iteration 88
Picard's method of successive approximations 88
pitchfork bifurcation 114
planar system 22
Poincaré–Bendixson theorem 115
positively invariant set 174
potential energy 56

R
radioactive decay 15
Rayleigh's equation 215
reduction of order 55
Riccati equation 14
Richardson model 24
right-hand side 5

S
saddle point 49
saturated solution 3
second-order equation 54
sensibility rate 20
separable variables 6
shooting method 208
shooting numerical method 210
SIR model 73
SIR model with vaccination 215
slope field 47
solution 3
solution curve 46
solution of a system 22
spectral radius 168
spiral point 51
spring-mass oscillator equation 53
stable cycle 116
stable equilibrium 52
stable focus 52
stable manifold 175
stable manifold theorem 176

stable node 52
stable solution 97
stable spiral point 52
stable subspace 174
stable system 99
static optimization 122
stationary solution 47
subcritical Hopf bifurcation 116
successive approximations 89
supercritical Hopf bifurcation 116
superposition 27
superposition principle 10

T
time series 46
total energy 56
trajectory 46
transition matrix 32

U
undetermined coefficients 141
undetermined coefficients method 62
uniformly asymptotically stable solution 98
uniformly stable solution 98
uniqueness 78
uniqueness of solutions 77
unstable equilibrium 52
unstable focus 52
unstable manifold 175
unstable node 52
unstable solution 97
unstable spiral point 52
unstable subspace 174

V
Van der Pol's equation 215
variation of parameters 11, 41, 61, 70
vector field 47
Verhulst model 19

W
Wronskian 67